Time Series Data Analysis Using EViews

Time Series Data Analysis Using EViews

Editor

Lavra Filipek

Time Series Data Analysis Using EViews

Edited by **Lavra Filipek**

Printed in 2017

ISBN: 978-1-68117-193-7

Library of Congress Control Number: 2015950791

© 2016 by
SCITUS Academics LLC,
616, Corporate Way, Suite 2, 4766,
Valley Cottage, NY 10989

www.scitusacademics.com

Preface

EViews (Econometric Views) is a statistical package for Windows, used mainly for time-series oriented econometric analysis. Basic time series modelling in EViews, including using lags, taking differences, introducing seasonality and trends, as well as testing for serial correlation, estimating ARIMA models, and using heteroskedastic and autocorrelated consistent standard errors. EViews can be applied for general statistical analysis and econometric analyses, such as cross-section and panel data analysis and time series estimation and forecasting. EViews combines spreadsheet and relational database technology with the traditional tasks found in statistical software, and uses a Windows GUI.

This book provides a hands-on practical guide to using the most suitable models for analysis of statistical data sets using EViews - an interactive Windows-based computer software program for sophisticated data analysis, regression, and forecasting - to define and test statistical hypotheses. Rich in examples and with an emphasis on how to develop acceptable statistical models, Time Series Data Analysis Using EViews presents statistical or econometric models for time series data. This book is designed as a reference tool to time series analysis in a very powerful and popular econometric software, EViews. It will also address the modules and structures of EViews that will help readers to fully harness the capabilities of the software.

Contents

A Multiplicative Seasonal ARIMA/ GARCH Model in EVN Traffic Prediction

Quang Thanh Tran[1], Zhihua Ma[2], Hengchao Li[1], Li Hao[1] and Quang Khai Trinh[3]

[1]Sichuan Provincial Key Laboratory of Information Coding and Transmission, Southwest Jiaotong University, Chengdu, China.
[2]Beijing Branch of China United Network Communications Co. Ltd., Beijing, China.
[3]Telecommunication Department, University of Transport and Communications, Hanoi, Vietnam.

ABSTRACT

This paper highlights the statistical procedure used in developing models that have the ability of capturing and forecasting the traffic of mobile communication network operating in Vietnam. To build such models, we follow Box-Jenkins method to construct a multiplicative seasonal ARIMA model to represent the mean component using the past values of traffic, then incorporate a GARCH model to represent its volatility. The traffic is collected from EVN Telecom mobile communication network. Diagnostic tests and examination of forecast accuracy measures indicate that the multiplicative seasonal ARIMA/GARCH model, i.e. ARIMA $(1, 0, 1) \times (0, 1, 1)_{24}$/GARCH $(1, 1)$ shows a good estimation when dealing with volatility clustering in the data series. This model can be considered to be a flexible model to capture well the characteristics of EVN traffic series and give reasonable forecasting results. Moreover, in such situations that the volatility is not necessary to be taken into account, i.e. short-term prediction, the multiplicative seasonal ARIMA/GARCH model still acts well with the GARCH parameters adjusted to GARCH $(0, 0)$.

INTRODUCTION

Traffic prediction is a key factor for a better network management which is now very important due to the explosive development of mobile communications and internet, especially in Vietnam, where there is a violent competition between so many service providers.

Statistical procedure has been used in developing forecasting models that have been applied to many different areas such as seasonal ARIMA in wireless traffic modeling and prediction [1]-[3], or ARCH in load forecasting [4]. Those analyses present many successful applications of ARIMA in forecasting time series data. However, ARIMA can only help presenting the conditional mean of the series. With the implicit assumption of homoscedasticity, GARCH is absolutely efficient in investigating the volatility characteristics of time series. Therefore, the combination of ARIMA and GARCH is a good choice to give a better result in capturing and forecasting time series such as wireless traffic data [5]-[7], crude oil prices data [8], inflation data [9], or internet traffic [10].

In this paper, the combination of ARIMA and GARCH is applied to mobile traffic in the condition of Vietnam, which has never been discussed before. A multiplicative seasonal ARIMA/GARCH model is built to fit and forecast EVN traffic. The evaluation of information criterion and forecast performance is made. The paper is organized as follows: Section 2 proposes to use multiplicative seasonal ARIMA/GARCH model to fit and forecast EVN traffic. Section 3 presents the experiment results and discussions. Finally, the conclusions are given in Section 4

PROPOSE TO BUILD A MULTIPLICATIVE SEASONAL ARIMA/ GARCH MODEL

The detail explanations of a multiplicative seasonal ARIMA model and a GARCH model can be found in references [1] [9] and [11]-[14], respectively. Below is the briefly description of a multiplicative seasonal ARIMA model which is derived from [1]:

$$\phi_p(B)\Phi_P(B^s)\nabla^d\nabla_s^D X_t = \theta_q(B)\Theta_Q(B^s)a_t \tag{1}$$

Or

$$W_t = \nabla^d\nabla_s^D X_t \tag{2}$$

where,

$$W_t = \phi_p^{-1}(B)\Phi_P^{-1}(B^s)\theta_q(B)\Theta_Q(B^s)a_t \tag{3}$$

To build a multiplicative seasonal ARIMA/GARCH model, we first construct a multiplicative seasonal ARIMA to present the mean component using the past values of the EVN traffic. We then incorporate a GARCH model to represent its volatility. The whole progress can be described in the flowchart in Figure 1 below. The progress in the flowchart can be expressed step by step as follow:

Step 1: Using spectrum analysis to determine the period s of the traffic trace

This step is very important to give a consideration to a seasonal ARIMA model. If s is found, then we can make a decision of a multiplicative seasonal ARIMA which is in the form of (p, d, q) × (P, D, Q).

Step 2: Identification of stationary, determine d and D

The second step is also very important in fitting an ARIMA model. It is the determination of the order of differencing needed to make the series stationary.

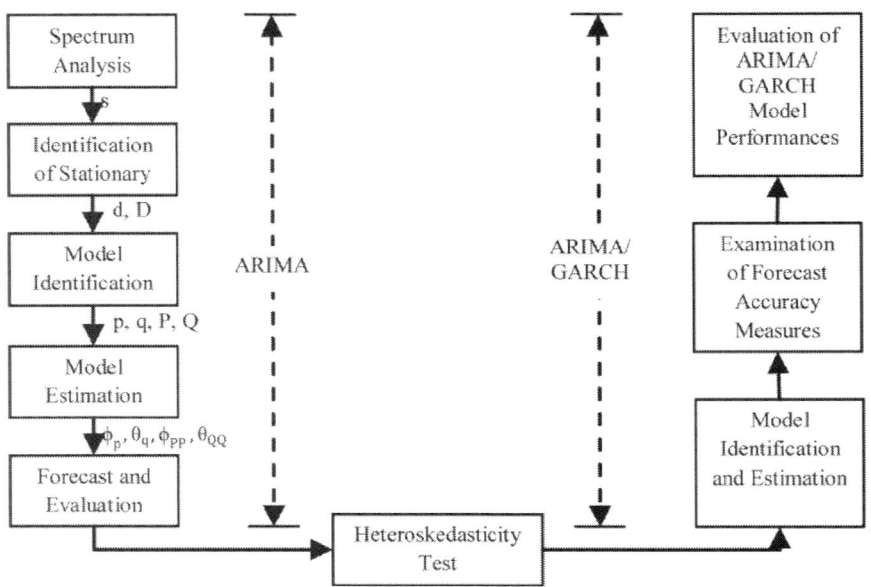

Figure 1: Flowchart of building ARIMA/GARCH process.

Step 3: Model identification, determining all the orders

Propose to begin with candidate parameter sets that have small (p, q) , (P,Q) values such as 0, 1, or 2 but where p, P and q, Q should not be 0 simultaneously in one set. Then, we can select the best (p, q) , (P,Q) combination according to the known model identification such as AIC (Akaike Information Criterion) and BIC (Bayesian Information Criterion) [15].

Step 4: Model estimation

Estimating all the parameters using approximate maximum likelihood parameter estimation methods, so that we obtain:

$$\phi_1, \phi_2, \cdots, \phi_p, \ \theta_1, \theta_2, \cdots, \theta_q, \ \phi_{P1}, \phi_{P2}, \cdots, \phi_{PP}, \ \theta_{Q1}, \theta_{Q2}, \cdots, \theta_{QQ}$$

Step 5: Forecast and evaluation

We use the fitted multiplicative seasonal ARIMA models obtained from (3) to forecast the series.

Step 6: Heteroskedasticity test

The existence of heteroscedasticity in hourly EVN traffic series must be examined before starting to estimate the GARCH model.

Step 7: Model identification and estimation for multiplicative seasonal ARIMA/GARCH model

We now verify the adequacy of AR and MA terms of the mean equation by implementing the correlogram Qtest, Jarque Bera test and ARCH test on the stationary series achieved from step 2.

The result of no serial correlation under the correlogram Q-test will indicate that we can proceed with the estimation of the conditional variance for the errors using GARCH. We limit the order of GARCH, (p, q) to 4, that is we use different orders of p, q = 0, 1, 2, 3 and 4, since GARCH is used for short-term forecasting. Incorporating the stationary series achieved from step 2 and the mean equation with AR and MA terms achieved from step 3, we estimate a GARCH model by finding a significant order combination under a specific error distribution (p-values should all be less than 0.10 level of significance and coefficient of the variance equation should all be positive).

Step 8: Examination of forecast accuracy measures

Static forecasting on the model is performed to show measures of fore-cast accuracy over the estimation period. The model with the smallest measure of forecast error will be chosen as the one with the most ac-curate fit of the time series model. Then, some more tests will be per-formed, such as correlogram of standardized residuals squared which consists of autocorrelation and partial auto-correlation, test for pre-senting of conditional heteroscedasticity in the data with ARCH-LM test on the residuals, standardized residuals.

Step 9: Evaluation of multiplicative seasonal ARIMA/GARCH model performances

The final step is to evaluate the forecast performances by our achieved multiplicative seasonal ARIMA/ GARCH model. The evaluation includes the information criterion, i.e. AIC and SIC values in the estimation stage, and forecast performances in the forecasting stage.

EXPERIMENTAL RESULTS AND DISCUSSIONS

Through spectrum analysis, we can figure out the periodicity of 24 hours or one day and we can say that s = 24 for this traffic. Then, the processes on the correlogram of EVN traffic stream show that the series needs to take the logarithm transformation (EVNLOG) and the 24-period seasonal difference (EVNLOGd0D1) to become variance stationary. Refer to the equation described in (2) we have:

$$W_t = \nabla^1 \nabla^1_{24} \ln X_t \tag{4}$$

where X_t is our series EVN, and W_t is EVNLOGd0D1.

In the next step, the estimation performed by EViews shows that the chosen model should have AR (1), MA (1) and SMA (24) components can be expressed as: ARIMA (1, 0, 1) × (0, 1, 1)$_{24}$ which is implemented on the logarithm form of the original series. Also, the coefficients are:

EVNLOGD0D1

$$= 0 + \left[AR(1) = 0.636844024029, MA(1) = 0.316609103164, SMA(24) = -0.9415532237619 \right]$$

The obtained fitted multiplicative seasonal ARIMA model can be expressed detail as:

$$\left(1 - \phi_1(B)\right) \nabla^1_{24} \ln X_t = \left(1 - \theta_1(B)\right)\left(1 - \Theta_1\left(B^{24}\right)\right) a_t \tag{5}$$

$$\Leftrightarrow \nabla^1_{24} \ln X_t - \phi_1 \nabla^1_{24} \ln X_{t-1} = \left(1 - \Theta_1\left(B^{24}\right)\right)\left(a_t - \theta_1 a_{t-1}\right) \tag{6}$$

$$\Leftrightarrow \nabla^1_{24} \ln X_t - \phi_1 \left(\ln X_{t-1} - \ln X_{t-25} \right) = a_t - \theta_1 a_{t-1} - \Theta_1 \left(B^{24} \right) a_t + \theta_1 \Theta_1 \left(B^{24} \right) a_{t-1} \tag{7}$$

$$\Leftrightarrow \left(\ln X_t - \ln X_{t-24} \right) - \phi_1 \left(\ln X_{t-1} - \ln X_{t-25} \right) = a_t - \theta_1 a_{t-1} - \Theta_1 \left(a_t - a_{t-24} \right) + \theta_1 \Theta_1 \left(a_{t-1} - a_{t-25} \right) \tag{8}$$

$$\Leftrightarrow \ln X_t = \phi_1 \ln X_{t-1} + \ln X_{t-24} - \phi_1 \ln X_{t-25} + (1 - \Theta_1) a_t - \theta_1 (1 - \Theta_1) a_{t-1} + \Theta_1 a_{t-24} + \theta_1 \Theta_1 a_{t-25} \tag{9}$$

where,

$\phi_1 = 0.6368$, $\theta_1 = 0.3166$, $\Theta_1 = 0.9416$

$$\Rightarrow \ln \hat{X}_t = \hat{\beta}_1 \ln \hat{X}_{t-1} + \hat{\beta}_2 \ln \hat{X}_{t-24} - \hat{\beta}_3 \ln \hat{X}_{t-25} - \hat{\theta}_1 \hat{a}_{t-1} + \hat{\theta}_2 \hat{a}_{t-24} + \hat{\theta}_3 \hat{a}_{t-25} \tag{10}$$

where, $\hat{a}_{-1} = \ln X_{t-1} - \ln \hat{X}_{t-1}$, $\ln X_{t-1}$ is actual values and $\ln \hat{X}_{t-1}$ is forecast values.

The forecast of EVN traffic stream using multiplicative seasonal ARIMA (1, 0, 1) × (0, 1, 1)$_{24}$ model is now conducted. EViews software provides the one-step ahead static forecasts which are more accurate than the dynamic forecasts. Static forecasting extends the forward recursion through the end of the estimation sample, allowing for a series of one-step ahead forecasts of both the structural model and the innovations. When computing static forecasts, EViews uses the entire estimation sample to backcast the innovations [16].

In Figure 2, the graph of actual hourly EVN traffic stream is plotted using a solid red line and while blue line represents the forecasted hourly EVN traffic stream by ARIMA (1, 0, 1) × (0, 1, 1)$_{24}$. The forecast series follow the actual series closely.

In the next step, the heteroscedasticity test is implemented and it shows that our traffic data contains volatility periods. Thus, we can proceed to build GARCH model based on the multiplicative seasonal

ARIMA model that we achieved. Following the steps mentioned above, GARCH (1, 1) assuming GED formulates which has the smallest measure of forecast error, i.e. MAE and RMSE, should be chosen as the one with the most accurate fit of the time series model. MAE indicates that the average difference between the forecast and the observed value of the model is 0.080042, while RMSE and MAPE are 0.131390 and 276.0843, respectively.

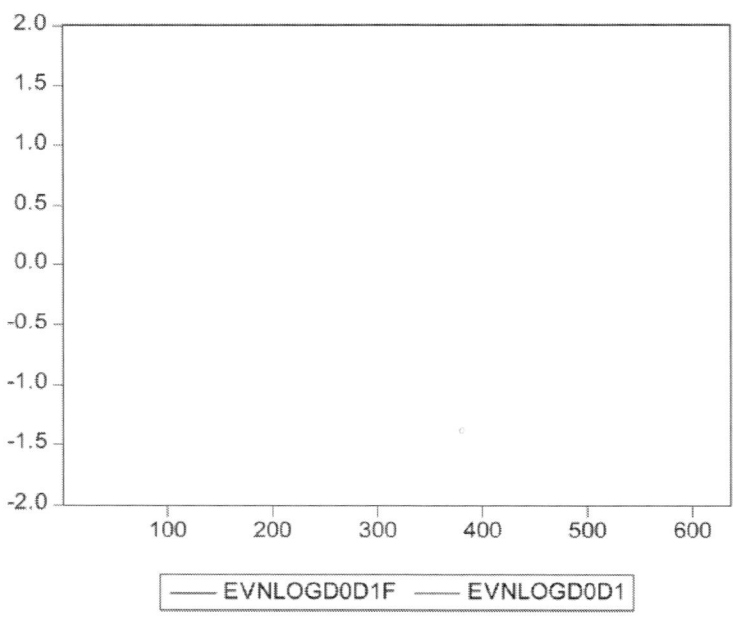

Figure 2: The plot of actual values and forecast values by ARIMA (1, 0, 1) × (0, 1, 1)$_{24}$ model.

Incorporating the most adequate choice for the volatility model, we now present the forecast for the mean and error variance of the EVN traffic, as shown in Figure 3 using the in-sample observations under static forecasting. The figure implies that volatile values are evident during the values between about 280 to 290 and 480 to 490. This is evident in the wide confidence intervals on the GARCH model under the forecast of mean. For the other values, however, we observe a stable and predictable traffic, as shown in the low values of the forecast of error variance.

Furthermore, some other tests are also implemented to make our decision more convincing. In this case, the correlogram of standardized residuals squared once again proves that the model is adequate, and the ARCH-LM test on the residuals of this model indicates that the conditional heteroscedasticity is no longer present in the data.

Next we plot the actual and forecast EVN traffic value by ARIMA (1, 0, 1) × (0, 1, 1)$_{24}$/GARCH (1, 1) model. From Figure 4, it can be concluded that the trend of forecast values follows the actual EVN traffic values closely.

In the final step, we will evaluate our ARIMA (1, 0, 1) ×(0, 1, 1)$_{24}$/ GARCH (1, 1) model in terms of AIC and SIC values in the estimation stage, and forecast performances in the forecasting stage.

- Information Criterion for ARIMA (1, 0, 1) × (0, 1, 1)$_{24}$/GARCH (1, 1) Models

In the model estimation step, the AIC and SIC values from ARIMA (1, 0, 1) × (0, 1, 1)$_{24}$/GARCH (1, 1) model is calculated. According to our criterion, the smaller AIC and SIC values, the better model defined. The results are tabulated in Table 1.

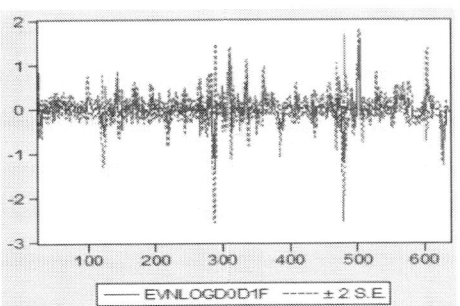 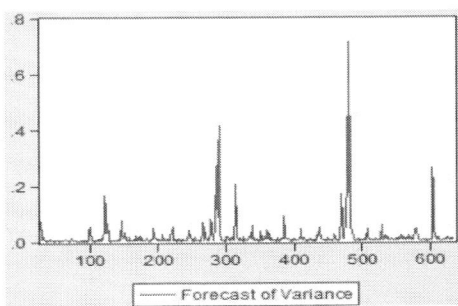

Figure 3: ARIMA (1, 0, 1) × (0, 1, 1)$_{24}$/GARCH (1, 1) model forecast for the mean and error variance of EVN traffic using the in-sample observations under static forecasting.

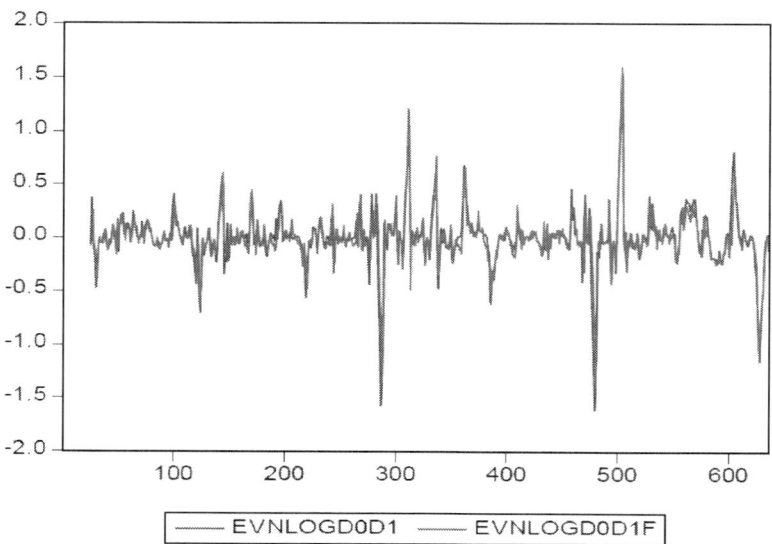

Figure 4: The plot of actual traffic against forecast traffic by ARIMA $(1, 0, 1) \times (0, 1, 1)_{24}$/GARCH $(1, 1)$ model.

In Table 1, AIC and SIC values obtained from the equation estimation of ARIMA $(1, 0, 1) \times (0, 1, 1)_{24}$/- GARCH $(1, 1)$ model using EViews are shown together with those of ARIMA $(1, 0, 1) \times (0, 1, 1)_{24}$ model which is a part of our ARIMA $(1, 0, 1) \times (0, 1, 1)_{24}$/GARCH $(1, 1)$ model. The AIC and SIC of ARIMA $(1, 0, 1) \times (0, 1, 1)_{24}$/GARCH $(1, 1)$ are -1.9226 and -1.8721, respectively, which can be found smaller than those of ARIMA $(1, 0, 1) \times (0, 1, 1)_{24}$ model, i.e.-1.2187 and -1.1970, respectively. This result shows that the GARCH part presents a positive influence and makes our ARIMA $(1, 0, 1) \times (0, 1, 1)_{24}$/GARCH $(1, 1)$ model built in a reasonable way.

- Forecasting Performance of ARIMA $(1, 0, 1) \times (0, 1, 1)_{24}$/GARCH $(1, 1)$ Model

The forecasting performance is evaluated via forecasting statistics that tabulated in **Figure** 5. The smaller those values are, the better forecasting performance obtained. The statistics in Figure 5 show that our ARIMA $(1, 0, 1) \times (0, 1, 1)24$/GARCH $(1, 1)$ model presents a reasonable result in forecasting EVN traffic series.

CONCLUSIONS

In our study, a multiplicative seasonal ARIMA/GARCH model, i.e. ARIMA $(1, 0, 1) \times (0, 1, 1)_{24}$/GARCH $(1, 1)$, shows a good result in describing and forecasting our mobile communication network traffic. The mobile traffic is found containing volatility periods. Therefore, the ARIMA $(1, 0, 1) \times (0, 1, 1)_{24}$ model is firstly formed to present the mean components, and then the GARCH $(1, 1)$ model is incorporated to deal with the volatility of the traffic. The evaluation of the estimation criterions shows that our ARIMA $(1, 0, 1) \times (0, 1, 1)_{24}$/GARCH $(1, 1)$ was built reason-ably with a significant impact of the GARCH part. And also the forecasting performance evaluation presents small forecasting error values that confirm the capable of fitting and forecasting the traffic of our model.

Moreover, as a part of our model, ARIMA $(1, 0, 1) \times (0, 1, 1)_{24}$ also present a relatively good result when conducting to fit and forecast the traffic. Based on this, we can conclude that in short-term prediction, where the volatility even occurs but has an insignificant impact on the whole result of forecast, our multiplicative seasonal ARIMA/GARCH model can be simplified as a multiplicative seasonal ARIMA model, by adjusting the parameter of the GARCH part, i.e. GARCH $(0, 0)$. In conclusion, our multiplicative seasonal ARIMA/GARCH model is a flexible model which is capable of fitting and forecasting mobile traffic not only in short-term prediction but also in long-term prediction.

ACKNOWLEDGMENTS

This work was supported in part by the Young Innovative Research Team of Sichuan Province under Grant 2011JTD0007, and by the Fundamental Research Funds for the Central Universities under Grant SWJTU12CX004, SWJTU12ZT02.

```
Forecast: EVNLOGD0D1F
Actual: EVNLOGD0D1
Forecast sample: 1 636
Adjusted sample: 26 636
Included observations: 611

Root Mean Squared Error        0.131390
Mean Absolute Error            0.080042
Mean Abs. Percent Error        276.0843
Theil Inequality Coefficient   0.253228
     Bias Proportion           0.000848
     Variance Proportion       0.046625
     Covariance Proportion     0.952527
```

Figure 5: Forecasting performances of ARIMA $(1, 0, 1) \times (0, 1, 1)_{24}$/ GARCH $(1, 1)$ model.

Table 1: Information criterion for ARIMA $(1, 0, 1) \times (0, 1, 1)_{24}$/ GARCH $(1, 1)$ and ARIMA $(1, 0, 1) \times (0, 1, 1)_{24}$ models

Model	AIC	SIC
ARIMA $(1, 0, 1) \times (0, 1, 1)_{24}$/ GARCH $(1, 1)$	**−1.9226**	−1.8721
ARIMA $(1, 0, 1) \times (0, 1, 1)_{24}$	−1.2187	−1.1970

REFERENCES

1. Shu, Y.T., Yu, M.F., Liu, J.K. and Yang, O.W.W. (2003) Wireless Traffic Modeling and Prediction Using Seasonal ARIMA Models. IEEE.
2. Yu, Y.H., Wang, J., Song, M.N. and Song, J.D. (2010) Network Traffic Prediction and Result Analysis Based on Seasonal ARIMA and Correlation Coefficient. 2010 International Conference on Intelligent System Design and Engineering Application. http://dx.doi.org/10.1109/ISDEA.2010.335
3. Guo, J., Peng, Y., Peng, X.Y., Chen, Q., Yu, J. and Dai, Y.F. (2009) Traffic Forecasting for Mobile Networks with Multiplicative Seasonal ARIMA Models. The Ninth International Conference on Electronic Measurement & Instruments.

4. Chen, H., Wan, Q.L., Zhang, B., Li, F.X. and Wang, Y.R. (2010) Short-Term Load Forecasting Based on Asymmetric ARCH Models. IEEE.

5. Chen, C.Y., Hu, J.M., Meng, Q. and Zhang, Y. (2011) Short-Time Traffic Flow Prediction with ARIMA-GARCH Model. IEEE Intelligent Vehicles Symposium (IV), Baden-Baden, 5-9 June.

6. Zhou, B., He, D. and Sun, Z.L. (2006) Traffic Predictability Based on ARIMA/GARCH Model. 2nd Conference on Next Generation Internet Design and Engineering.

7. Radha, S. and Thenmozhi, M. (2006) Forecasting Short Term Interest Rates Using ARMA, ARMA-GARCH and ARMA-EGARCH Models.

8. Nian, L.C. (2009) Application of ARIMA and GARCH Models in Forecasting Crude Oil Prices. A Dissertation Submitted in Partial Fulfillment of the Requirement for the Award of the Degree of Master of Science (Mathematics).

9. Ramon, H.L. (2008) Forecasting the Volatility of Philippine Inflation Using GARCH Models. BSP Working Paper Series, Series No. 2008-01.

10. Kim, S. (2011) Forecasting Internet Traffic by Using Seasonal GARCH Models. Journal of Communications and Networks, 13. http://dx.doi.org/10.1109/JCN.2011.6157478

11. Dong, N.Q. (2008) Econometrics—Advanced Program. Science and Technique Publisher.

12. Dong, N.Q. (2008) Econometrics Assignments—With EVIEWS. Science and Technique Publisher.

13. Quynh, N.H. (2004) Time Series Analysis and Identification. Science and Technique Publisher.

14. Agung, I.G.N. (2009) Time Series Data Analysis Using EViews. John Wiley & Sons (Asia) Pte Ltd.

15. (2007) EViews 6 User's Guide I. Quantitative Micro Software, LLC.

16. (2007) EViews 6 User's Guide II. Quantitative Micro Software, LLC.

CITATION

17. Quang Thanh Tran, Zhihua Ma, Hengchao Li, Li Hao and Quang Khai Trinh A Multiplicative Seasonal ARIMA/GARCH Model in EVN Traffic Prediction DOI: 10.4236/ijcns.2015.84005.

Simulation Analysis of Threshold Autoregressive Unit Root Tests

Steve Cook
Department of Economics, Swansea University,
Singleton Park, Swansea SA2 8PP, UK

2

ABSTRACT

Using numerical simulation, the finite-sample properties of threshold autoregressive (TAR) and momentum-threshold (MTAR) autoregressive-based unit root tests under both deterministic and consistent methods of threshold estimation are examined in the presence of generalised autoregressive conditional heteroskedasticity (GARCH). Previous research is extended by considering both the impact of alternative robust methods of covariance matrix estimation and the behaviour of the secondary tests of asymmetry associated with the TAR and MTAR models. The results obtained reveal many interesting features, in particular the distortionary effects of consistent-threshold estimation. In summary, the findings presented indicate that caution should be exercised when interpreting the results of these frequently employed threshold-based testing methods.

INTRODUCTION

Following the work of Engle [1], Bollerslev [2], and Taylor [3], the analysis of volatility has become a central feature of both empirical and theoretical finance. This in turn has led to the emergence of a large literature examining the properties and behaviour of unit root tests in the presence of volatility in its commonly considered form of generalised autoregressive conditional heteroskedasticity (GARCH). (The GARCH model has become a cornerstone of empirical research in economics and finance, receiving widespread application and proving to provide accurate forecasts (see [4]).) While some authors have considered joint estimation of unit root testing equations and GARCH (1, 1) processes (see, inter alia, [5–7] and references contained therein), others have examined the size properties of unit root tests in the presence of neglected GARCH behaviour (see [8–10]). (Following the emergence of a number of studies considering the relevance and implications of heavy-tailed distributions in economic and financial time series data (see, inter alia, [11–14]), a further related literature has emerged examining the properties of unit root tests when applied to series with heavy-tailed disturbances (see [15, 16]).) In this paper, this latter literature is extended by considering the properties of the threshold autoregressive (TAR) and momentum-threshold autoregressive (MTAR) unit root tests of Enders and Granger [17] when applied to time series with GARCH (1,1) disturbances. This issue clearly merits attention given the noted prevalence of GARCH behaviour in economic and financial data. While Cook [9] provides an initial examination of the properties of the TAR and MTAR unit root tests in the presence of GARCH, the current study develops this work in two crucial ways. First, the impact of alternative methods of variance-covariance estimation is considered. An analysis of this is of clear importance as while the covariance matrix estimators of White [18] and Newey and West [19] are routinely adopted by econometricians and time series analysts as a general correction for problems encountered with heteroskedasticity, their impact in the presently considered circumstances is unknown. Second, the subsequent tests of the hypothesis of symmetry associated with the TAR and MTAR tests are examined. Again, this is an important issue as the results of these subsequent tests of symmetry are often reported by investigators using Fisher's F-distribution to reinforce their empirical analyses, despite the individual properties of the tests in these circumstances, and the applicability of the F-distribution, being unknown. (The use of the F-distribution to provide critical values for the symmetry tests is apparent in the empirical analysis conducted in the seminal work of Enders and Granger [17]. More recent studies employing the F-distribution include, inter alia, Payne and Moham-madi [20] and Thompson [21].) Indeed, it will be seen that the use of the *F*-distribution can generate misleading results for the symmetry tests, irrespective of whether GARCH is present. The present work ad-dresses these issues using Monte Carlo simulation, considering the TAR and MTAR tests under both deterministic and consistent methods of threshold estimation.

This paper proceeds as follows. In the following section the TAR and MTAR tests are outlined. Using Monte Carlo methods, Section 3 con-siders the finite-sample distributions of the TAR and MTAR unit root tests under alternative methods of covariance matrix estimation and the empirical sizes of the symmetry tests under use of the *F*-distribution. The simulation analysis is extended in Section 4 to consider the behav-iour of the TAR and MTAR unit root tests and the subsequent tests of symmetry, in the presence of GARCH. Section 5 concludes.

TAR AND MTAR UNIT ROOT TESTS

Consider a time series process $\{y_t\}_{t=0}^T$ Given this series, the familiar Dickey-Fuller [22] (DF) test in its simplest form can be expressed as the t-ratio of $\hat{\phi}$ in the following regression:

$$\Delta y_t = \phi y_{t-1} + \varepsilon_t.$$

$$(2.1)$$

Comparison of the calculated test statistic with specifically derived nonstandard critical values allows examination of the unit root hypoth-esis $H_0:\phi=0$ against an alternative of asymptotic stationarity ($H_0:\phi<0$). However it is apparent that (2.1) is an implicitly symmetric specifica-tion. To allow for the possibility of stationary asymmetric adjustment about an underlying attractor, Enders and Granger [17] draw upon the threshold autoregressive methods of Tong [23, 24]. Adopting this approach, (2.1) is generalised via the introduction of the Heaviside indicator function to partition the lagged level term y_{t-1} with a resulting testing equation given as

$$\Delta y_t = I_t \rho_1 y_{t-1} + (1 - I_t)\rho_2 y_{t-1} + \xi_t,$$

$$(2.2)$$

where the zero-one Heaviside indicator function (I_t) can be specified as either

$$I_t = \begin{cases} 1 & \text{if } y_{t-1} \geq 0, \\ 0 & \text{if } y_{t-1} < 0 \end{cases}$$

$$(2.3)$$

or:

$$I_t = \begin{cases} 1 & \text{if } \Delta y_{t-1} \geq 0, \\ 0 & \text{if } \Delta y_{t-1} < 0. \end{cases}$$

$$(2.4)$$

The combination of (2.2) and (2.3) results in a TAR model, while combination of (2.2) and (2.4) leads to an MTAR model. The testing equation (2.2) therefore provides the basis of the asymmetric unit root tests of Enders and Granger [17], with the speed of adjustment about the stationary attractor given by ρ_1 when above equilibrium and ρ_2 when below. The unit root hypothesis is then tested via the null $H_0 : \rho_1 = \rho_2 = 0$, with the resulting test statistic denoted as Φ when using (2.3) and Φ^* when using (2.4). Due to the nonstandard distributions of the Φ and Φ^* statistics, their use requires the derivation of appropriate critical values via Monte Carlo experimentation. To further examine potential asymmetric stationarity, investigators often conduct a further test of the symmetry hypothesis via the null $\rho_1 = \rho_2$ using a conventional

F-statistic. The appropriateness of the F-distribution when conducting this subsequent test of symmetry will be questioned later in this paper.

The above analysis adopts an implicit assumption of zero attractor ($y_t=0$). However, in practice it is more realistic to consider a nonzero attractor. In these circumstances, it is possible to derive the required threshold either deterministically or via consistent-threshold estimation. Under the first option, a deterministic threshold is imposed via regressing the original series $\{y_t\}$ upon a constant to derive a new series $\{\tilde{y}_t\}$. (In this analysis, the intercept-only model is considered for the TAR and MTAR tests as it is this form alone which is available for the tests under consistent-threshold estimation.) The derived series $\{\tilde{y}_t\}$ is then employed in a modified version of (2.2) as given below:

$$\Delta\tilde{y}_t = I_t\rho_1\tilde{y}_{t-1} + (1 - I_t)\rho_2\tilde{y}_{t-1} + \zeta_t.$$

$$(2.5)$$

In a similar manner, the specification of the Heaviside indicator function in (2.3) or (2.4) is modified by using the appropriate revised series $\{\tilde{y}_t\}$ rather than $\{y_t\}$. The resulting TAR and MTAR tests are denoted as Φ_μ and Φ_μ^*, respectively. Under consistent-threshold estimation, the above two-step procedure is replaced by an optimisation procedure to select the threshold. To illustrate this, consider the Φ_μ and Φ_μ^* tests above. To construct these tests statistics, prior to regression upon a constant term effectively, this demeans the series of interest, implying that the version of (2.2) actually estimated is given as

$$\Delta y_t = I_t\rho_1(y_{t-1} - \overline{y}) + (1 - I_t)\rho_2(y_{t-1} - \overline{y}) + \xi_t.$$

$$(2.6)$$

However, as noted by Enders [25], in the presence of asymmetric adjustment $\rho1$ and $\rho2$ differ, and consequently the mean is a biased estimator of the threshold. To overcome this and obtain a consistent value of the threshold, the approach of Chan [26] is followed with a grid search procedure employed to select the threshold. The previously defined Heaviside indicator functions of (2.3) and (2.4) are then revised as follows:

$$
I_t = \begin{cases} 1 & \text{if } y_{t-1} \geq \tau, \\ 0 & \text{if } y_{t-1} < \tau, \end{cases}
$$

(2.7)

$$
I_t = \begin{cases} 1 & \text{if } \Delta y_{t-1} \geq \tau\Delta, \\ 0 & \text{if } \Delta y_{t-1} < \tau\Delta, \end{cases}
$$

(2.8)

with the thresholds τ and τ_Δ selected by grid search over a range of values of $\{yt\}$ and $\{\Delta yt\}$ respectively. The appropriate asymmetric unit root testing equations are therefore given as

$$
\Delta y_t = I_t \rho_1 (y_{t-1} - \psi) + (1 - I_t)\rho_2 (y_{t-1} - \psi) + \eta_t, \quad \psi = \tau, \tau_\Delta.
$$

(2.9)

Application of the consistent-threshold versions of the TAR and MTAR tests therefore requires selection of the appropriate threshold value (ψ). Considering TAR adjustment, the series of interest $\{y_t\}$ is reordered as y_1^0 $<y_2^0 <\cdots<y_T^0$, with the central 70% of observations (y_i^0,=$0.15T$,...,$0.85T$) within this range of values considered as potential thresholds. Each of these values ($y_i^0 = \tau$) is employed in turn in the indicator function of (2.7) with (2.9) then estimated. The value y_i^0 delivering the minimum residual sum of squares $(\Sigma \eta_t^2)$ for (2.9) is then deemed to be the consistent estimate of the threshold (τ) with the resulting TAR asymmetric unit root test labelled Φ_μ^c. Under MTAR adjustment, a similar approach is followed with the central 70% of observations from the reordered sequence $\{\Delta y_1^0, \Delta y_2^0, ..., \Delta y_T^0\}$ now considered as potential thresholds. Again the selected threshold $\tau\Delta$ is that value minimising the residual

sum of squares of (2.9) with (2.8) now employed as the appropriate indicator function. The resultant MTAR test under consistent-threshold estimation is labelled as $\Phi_\mu^{*,c}$. In the subsequent research of Cook and Manning [27], the use of consistent-threshold estimation was shown to substantially increase the power of both TAR and MTAR asymmetric unit root tests in the presence of asymmetrically stationary processes. In the subsequent sections of this paper, the behaviour of the above tests will be examined in the presence of GARCH, with particular attention paid to the role of alternative variance-covariance matrix estimators and the secondary test of symmetry.

MONTE CARLO EXPERIMENTATION I

Derivation of Critical Values for the Asymmetric Unit Root Tests

Before considering the properties of the Φ_μ, Φ_μ^*, Φ_μ^c, and $\Phi_\mu^{*,c}$ tests and their associated subsequent tests of symmetry in the presence of GARCH, critical values for these tests are derived under the use of alternative variance-covariance estimators. In addition to considering the standard OLS covariance estimator, the covariance matrix estimators of White [18] and Newey-West [19] are also employed for the four tests. The use of White and Newey-West covariance matrix estimators is denoted by the addition of the subscripts w and n, respectively, to the test statistic, while no additional subscript is employed when using the standard OLS covariance estimator. To obtain the required critical values, the following data generation process (DGP) is employed:

$$y_t = y_{t-1} + \varepsilon_t \quad t = 1, \ldots, T,$$

(3.1)

where the innovation series $\{\varepsilon_t\}$ is generated using pseudo i.i.d. N(0,1) random numbers from the NRND procedure in the EViews 6. The experiments are performed over 50,000 simulations for three sample sizes: $T=\{200,400,800\}$. Critical values are reported in Table 1 for three levels of significance ($\alpha=0.10,0.05,0.01$). The consideration of four tests, three sample sizes, three covariance matrix estimators, and three levels of significance results in the calculation of ($4\times3\times3\times3=$) 108 critical values. For all of the four tests examined, the results presented in Table 1 show that finite-sample critical values are clearly dependent upon the covariance matrix estimator employed. Although the impact of alternative covariance estimators is negligible asymptotically, its finite-sample impact is apparent and is noticeable for even the relatively large sample of 800 observations considered in the present experiments. In short, it can be seen that movement from the OLS covariance matrix estimator to the White estimator and then Newey-West estimator leads to inflation of the critical values of all tests.

Table 1: Critical values for asymmetric unit root tests

$\Phi_{\mu,w}^{*,c}$	4.87	5.98	8.61	4.60	5.60	7.78	4.49	5.38	7.48
$\Phi_{\mu,n}^{*,c}$	5.63	7.02	10.47	5.12	6.33	9.06	4.79	5.85	8.20

The reported results represent critical values for the TAR and MTAR tests under deterministic and consistent threshold estimation using alternative covariance matrix estimators. The results were derived over 50,000 simulations.

Empirical Sizes of the Symmetry Tests

To consider the finite-sample sizes of the subsequent associated tests of symmetry ($H_0:\rho_1=\rho_2$) associated with the Φ_μ, Φ_μ^*, Φ_μ^c, and $\Phi_\mu^{*,c}$ tests under alternative methods of covariance matrix estimation, the DGP of the previous subsection is employed but with (3.1) respecified as follows:

$$y_t = \rho y_{t-1} + \varepsilon_t \quad t = 1,\ldots,T.$$

$$(3.2)$$

This modification is introduced to allow consideration of stationary processes via $|\rho|<1$, as the symmetry test is valid for stationary processes only. To examine the finite-sample empirical sizes of the symmetry tests associated with the Φ_μ, Φ_μ^*, Φ_μ^c, and $\Phi_\mu^{*,c}$ statistics, $\rho=0.85$ is employed in the following Monte Carlo analysis. The empirical rejection frequencies for the symmetry tests are derived at the 5% level of significance ($\alpha=0.05$) over 50,000 replications with the innovation series $\{\varepsilon_t\}$ again generated as a standard normal variable using the NRND procedure.

The empirical rejection frequencies for the symmetry tests are reported in Table 2. From inspection of this table, it is clear that there are a number of interesting findings to consider. First, under deterministic estimation of the threshold, the symmetry test for the TAR model is substantially undersized irrespective of the covariance matrix estimator employed. In contrast to this, the symmetry test for the MTAR model with a deterministic threshold has near nominal size. Second, under consistent-threshold estimation, the symmetry tests for both the TAR and MTAR models are dramatically oversized, with the former exhibiting greater size distortion than the latter. Third, movement from the standard OLS covariance matrix estimator to the White and the Newey-West estimator results in an increase in empirical size irrespective of the test and method of threshold estimation employed.

Table 2: Empirical sizes of symmetry tests

	$T = 200$	$T = 400$	$T = 800$
Φ_μ	0.08	0.09	0.12
$\Phi_{\mu,w}$	0.12	0.10	0.13
$\Phi_{\mu,n}$	0.22	0.19	0.16
Φ_μ^c	33.15	29.42	26.70

	$T = 200$	$T = 400$	$T = 800$
$\Phi^{c}_{\mu,w}$	34.78	30.07	26.99
$\Phi^{c}_{\mu,n}$	36.02	31.33	27.73
Φ^{*}_{μ}	4.35	4.45	4.39
$\Phi^{*}_{\mu,w}$	5.00	4.70	4.60
$\Phi^{*}_{\mu,n}$	6.16	5.46	5.07
$\Phi^{*,c}_{\mu}$	26.69	24.37	20.69
$\Phi^{*,c}_{\mu,w}$	29.94	25.68	21.21
$\Phi^{*,c}_{\mu,n}$	33.33	27.77	22.43

The reported results represent empirical rejection frequencies at the 5% level of significance of symmetry tests for the TAR and MTAR models under deterministic and consistent threshold estimation using alternative covariance matrix estimators. The results were derived over 50,000 simulations.

MONTE CARLO EXPERIMENTATION II

Empirical Sizes of Asymmetric Unit Root Tests in the Presence of GARCH

To examine the properties of the above-mentioned asymmetric unit root tests in the presence of GARCH (1,1) errors, the following DGP is employed:

$$y_t = y_{t-1} + w_t \quad t = 1, \ldots, T,$$

$$h_t^2 = \phi_0 + \phi_1 w_{t-1}^2 + \phi_2 h_{t-1}^2,$$

$$w_t = h_t v_t,$$

$$v_t \sim N(0, 1).$$

$$(4.1)$$

This DGP closely follows that of Kim and Schmidt [8] and considers GARCH processes generated for a range of realistic values of $\{\phi 1_{,2}\}$ corresponding to near integration, with $\phi_0=1-\phi_1-\phi_2$ in all cases. (In this paper the less empirically realistic, or relevant, cases of degenerate GARCH ($\phi_0=0$) and integrated GARCH ($\phi_1+\phi_2=1$) are not considered.) The precise values of the volatility parameter (ϕ_1) and the degree of persistence ($\phi_1+\phi_2$) employed are based upon estimated values observed in empirical research, with the distinction between typical values of $\{\phi_{1,2}\}$ observed at differing data frequencies noted (see [28, 29]). Throughout, the initial value of the conditional variance is set equal to one ($h_0=1$), with an additional, initial 400 observations of the generation of the GARCH process discarded prior to generation of the series of interest yt to remove the impact of the initial condition. The initial value of yt is set to zero ($y_0=0$) without loss of generality. The innovation series $\{v_t\}$ is generated using pseudo i.i.d. $N(0,1)$ random numbers from the NRND procedure in EViews 6 with all experiments performed over 20,000 simulations for the previously considered sample sizes: $T=\{200,400,800\}$. The empirical sizes of the alternative tests are derived as the rejection frequencies for the alternative tests at the 5% level of significance under the use of the OLS, White and Newey-West covariance estimators using the critical values derived in the previous section.

Considering the results presented in Table 3 for the TAR unit root test for the standard OLS covariance estimator, it can be seen that the presence of GARCH does result in size inflation, with this becoming more noticeable for larger values of the volatility parameter. From inspection of the results for $\{\phi_1,\phi_2\}=\{0.15,0.84\}$ and $\{\phi_1,\phi_2\}=\{0.25,0.70\}$ it is also apparent that the degree of near integration as well as the size of the volatility parameter influences size distortion, as empirical size is lower for the second pairing of parameters with a lower sum for the GARCH parameters, but a higher degree of volatility. While these results hold under both deterministic and consistent threshold estimation, it can be seen that greater size distortion results from the use of consistent-threshold estimation. Conversely, it is apparent that the use of either the White or Newey-West covariance covariance matrix estimators dramatically reduces oversizing for the tests to a near negligible level.

Table 3: TAR unit root testing in the presence of GARCH

		(ϕ_1, ϕ_2)				
		(0.01, 0.98)	(0.05, 0.94)	(0.10, 0.89)	(0.15, 0.84)	(0.25, 0.70)
$T = 200$	Φ_μ	5.24	6.27	7.79	8.94	8.51
	$\Phi_{\mu,w}$	5.07	5.22	5.23	5.15	4.95
	$\Phi_{\mu,n}$	5.26	5.23	5.18	4.91	4.38
	Φ_μ^c	5.32	6.47	8.76	10.83	12.25
	$\Phi_{\mu,w}^c$	4.97	5.16	5.20	5.23	5.39
	$\Phi_{\mu,n}^c$	5.02	5.10	5.24	5.24	5.11
$T = 400$	Φ_μ	5.28	6.31	8.00	9.25	8.29
	$\Phi_{\mu,w}$	5.15	5.18	5.22	5.00	4.84
	$\Phi_{\mu,n}$	5.24	5.25	5.31	5.04	4.38
	Φ_μ^c	5.42	7.09	10.38	13.15	13.49
	$\Phi_{\mu,w}^c$	5.28	5.27	4.95	4.65	4.60
	$\Phi_{\mu,n}^c$	5.37	5.30	5.12	4.64	4.44
$T = 800$	Φ_μ	5.02	5.99	7.42	8.94	7.62
	$\Phi_{\mu,w}$	4.85	4.80	4.69	4.67	4.90
	$\Phi_{\mu,n}$	4.92	4.86	4.91	4.77	4.63
	Φ_μ^c	5.16	7.13	10.53	14.11	12.89
	$\Phi_{\mu,w}^c$	4.93	4.86	4.34	4.32	4.41
	$\Phi_{\mu,n}^c$	4.94	4.82	4.59	4.28	3.75

The reported results represent empirical sizes of the test of the unit root hypothesis at the 5% level of significance for the TAR model under deterministic and consistent threshold estimation using alternative covariance matrix estimators. The results were derived over 20,000 simulations.

From inspection of the results for the MTAR unit root test in Table 4, it can be seen that the behaviour of the MTAR test closely mimics that of the TAR test. However, it is also apparent that under consistent-threshold estimation and use of the standard OLS covariance matrix estimator, the oversizing of the MTAR test is less than that of TAR test for the identical values of the GARCH parameters. However, despite this, empirical sizes more than twice the nominal size are obtained.

Table 4: MTAR unit root testing in the presence of GARCH

| | | (ϕ_1, ϕ_2) | | | | |
		$(0.01, 0.98)$	$(0.05, 0.94)$	$(0.10, 0.89)$	$(0.15, 0.84)$	$(0.25, 0.70)$
	Φ^*_{μ}	4.99	5.99	7.61	9.09	8.78
	$\Phi^*_{\mu,w}$	4.98	5.07	4.92	4.90	4.74
$T = 200$	$\Phi^*_{\mu,n}$	5.06	5.23	5.22	5.01	4.50
	$\Phi^{*,c}_{\mu}$	5.30	6.33	7.83	9.13	9.66
	$\Phi^{*,c}_{\mu,w}$	4.94	4.97	4.59	4.26	3.95
	$\Phi^{*,c}_{\mu,n}$	4.94	5.07	4.81	4.41	3.83
	Φ^*_{μ}	5.21	6.21	8.31	9.84	8.59
	$\Phi^*_{\mu,w}$	5.06	4.92	4.78	4.63	4.28
$T = 400$	$\Phi^*_{\mu,n}$	4.96	4.92	4.82	4.68	4.05
	$\Phi^{*,c}_{\mu}$	4.96	6.54	8.42	10.12	9.74
	$\Phi^{*,c}_{\mu,w}$	5.00	4.63	4.21	3.80	3.43
	$\Phi^{*,c}_{\mu,n}$	4.95	4.52	4.09	3.68	3.14
	Φ^*_{μ}	5.09	6.29	8.15	10.04	8.04
	$\Phi^*_{\mu,w}$	5.03	4.76	4.61	4.41	4.48

	(ϕ_1, ϕ_2)				
	(0.01, 0.98)	(0.05, 0.94)	(0.10, 0.89)	(0.15, 0.84)	(0.25, 0.70)
$T = 800$ $\quad \Phi^*_{\mu,n}$	5.25	5.02	4.82	4.59	4.43
$\Phi^{*,c}_{\mu}$	5.21	6.63	8.94	10.56	10.05
$\Phi^{*,c}_{\mu,w}$	5.21	4.88	4.27	3.60	3.26
$\Phi^{*,c}_{\mu,n}$	5.04	4.72	4.09	3.59	3.24

The reported results represent empirical sizes of the test of the unit root hypothesis at the 5% level of significance for the MTAR model under deterministic and consistent threshold estimation using alternative covariance matrix estimators. The results were derived over 20,000 simulations.

Empirical Sizes of the Symmetry Tests in the Presence of GARCH

In Tables 5 and 6, the empirical sizes of the symmetry tests associated with the TAR and MTAR unit root tests are reported. These results are obtained using the following DGP:

$$y_t = \rho y_{t-1} + w_t \quad t = 1, \ldots, T,$$

$$h_t^2 = \phi_0 + \phi_1 w_{t-1}^2 + \phi_2 h_{t-1}^2,$$

$$w_t = h_t v_t,$$

$$v_t \sim N(0,1), \tag{4.2}$$

where $\rho=0.85$ in (4.2). Aside from the consideration of stationary time series via the use of the AR(1)specification $y_t=0.85y_t-1+w_t$, the Monte Carlo analysis conducted in this subsection follows that of the previous subsection in terms of generation and replication. While a range of results is reported in Tables 5 and6, the findings can be summarised straightforwardly.

Table 5: TAR symmetry testing in the presence of GARCH

		(ϕ_1, ϕ_2)				
		(0.01, 0.98)	(0.05, 0.94)	(0.10, 0.89)	(0.15, 0.84)	(0.25, 0.70)
$T = 200$	Φ_μ	0.09	0.17	0.36	0.63	0.99
	$\Phi_{\mu,w}$	0.13	0.16	0.18	0.20	0.20
	$\Phi_{\mu,n}$	0.28	0.26	0.37	0.44	0.42
	Φ_μ^c	34.01	35.87	39.03	41.37	44.73
	$\Phi_{\mu,w}^c$	35.52	35.39	35.05	34.37	34.90
	$\Phi_{\mu,n}^c$	36.59	37.04	37.84	38.27	39.25
$T = 400$	Φ_μ	0.14	0.25	0.90	1.86	2.61
	$\Phi_{\mu,w}$	0.15	0.15	0.21	0.24	0.34
	$\Phi_{\mu,n}$	0.22	0.26	0.34	0.45	0.57
	Φ_μ^c	29.37	31.84	35.54	38.72	41.60
	$\Phi_{\mu,w}^c$	29.82	29.41	28.71	28.05	28.82
	$\Phi_{\mu,n}^c$	30.85	31.14	31.46	31.90	33.18
$T = 800$	Φ_μ	0.11	0.39	2.09	4.76	5.38
	$\Phi_{\mu,w}$	0.09	0.13	0.22	0.39	0.44
	$\Phi_{\mu,n}$	0.11	0.15	0.37	0.63	0.67
	Φ_μ^c	26.83	29.26	33.95	38.42	41.23

	(ϕ_1, ϕ_2)				
	(0.01, 0.98)	(0.05, 0.94)	(0.10, 0.89)	(0.15, 0.84)	(0.25, 0.70)
$\Phi^c_{\mu,w}$	26.52	25.20	23.73	23.01	24.20
$\Phi^c_{\mu,n}$	27.51	26.93	26.56	27.21	28.85

The reported results represent empirical sizes of the test of the symmetry hypothesis at the 5% level of significance for the TAR model under deterministic and consistent threshold estimation using alternative covariance matrix estimators. The results were derived over 20,000 simulations.

Table 6: MTAR symmetry testing in the presence of GARCH

		(ϕ_1, ϕ_2)				
		(0.01, 0.98)	(0.05, 0.94)	(0.10, 0.89)	(0.15, 0.84)	(0.25, 0.70)
	Φ^*_{μ}	4.35	5.41	7.43	9.80	11.62
	$\Phi^*_{\mu,w}$	4.98	4.94	5.29	5.42	5.46
$T = 200$	$\Phi^*_{\mu,n}$	6.11	6.22	6.71	7.23	7.43
	$\Phi^{*,c}_{\mu}$	26.71	30.73	35.95	40.28	43.13
	$\Phi^{*,c}_{\mu,w}$	29.73	30.09	29.37	28.93	27.49
	$\Phi^{*,c}_{\mu,n}$	33.15	33.93	33.98	34.06	33.42
	Φ^*_{μ}	4.50	6.12	9.35	12.64	14.41
	$\Phi^*_{\mu,w}$	4.68	4.82	4.90	5.09	5.10
$T = 400$	$\Phi^*_{\mu,n}$	5.68	5.87	6.18	6.55	6.56

		(ϕ_1, ϕ_2)				
		(0.01, 0.98)	(0.05, 0.94)	(0.10, 0.89)	(0.15, 0.84)	(0.25, 0.70)
	$\Phi_\mu^{*,c}$	25.11	30.24	37.19	43.18	44.91
	$\Phi_{\mu,w}^{*,c}$	25.72	26.05	25.66	24.81	23.72
	$\Phi_{\mu,n}^{*,c}$	28.06	28.62	29.15	29.14	28.95
	Φ_μ^{*}	4.73	7.14	12.12	17.05	18.63
	$\Phi_{\mu,w}^{*}$	4.68	4.94	5.04	5.16	4.94
$T = 800$	$\Phi_{\mu,n}^{*}$	5.20	5.40	5.84	6.30	6.29
	$\Phi_\mu^{*,c}$	21.68	27.70	37.28	44.45	44.76
	$\Phi_{\mu,w}^{*,c}$	21.49	21.27	21.02	20.70	20.04
	$\Phi_{\mu,n}^{*,c}$	22.60	23.19	24.03	24.32	24.09

The reported results represent empirical sizes of the test of the symmetry hypothesis at the 5% level of significance for the MTAR model under deterministic and consistent threshold estimation using alternative covariance matrix estimators. The results were derived over 20,000 simulations.

Considering the TAR model with a deterministically imposed threshold value, substantial undersizing is apparent. Conversely, when consistent-threshold estimation is employed, the resulting symmetry test is massively oversized, with the use of OLS covariance matrix inducing greater size inflation than the corrected covariance matrix estimators of White and Newey-West. Despite undersizing (oversizing) of the test with a deterministic (consistent) threshold being noted previously in the absence of GARCH, it is apparent that neglected GARCH behaviour does increase size distortion with size inflation being positively related to the degree of volatility as given by ϕ_1. Turning to the results for the MTAR test, under the use of a deterministic threshold and corrected covariance matrix estimator, the test has relatively good size, particularly when the White covariance matrix is used. However, when the standard OLS covariance is employed, substantial size distortion is exhibited for volatile GARCH processes. Considering the symmetry test for the MTAR model under consistent threshold estimation, size distortion is apparent irrespective of the covariance matrix estimator employed, although the corrected covariance estimators, and the White covariance matrix estimator in particular, do reduce this. The results obtained therefore indicate that use of the preferred method of consistent-threshold estimation results in substantial oversizing of the symmetry test for both the TAR and MTAR models with all covariance matrix estimators, leading to a spurious detection of asymmetric behaviour.

CONCLUDING REMARKS

The popularity of threshold-based unit root tests as a means of detecting asymmetric stationarity in time series processes has increased in recent years, with their analysis and application now widespread. This paper has examined the practically important issue of the finite-sample size properties of these TAR and MTAR tests in the presence of GARCH behaviour, with particular attention paid to the use of alternative covariance matrix estimators and the subsequent test of the symmetry hypothesis. The results of the Monte Carlo analysis undertaken show the TAR and MTAR tests to be oversized when examining the unit root hypothesis in the presence of GARCH, unless a corrected covariance matrix estimator (White or Newey-West) is employed. This result is particularly apparent when considering more highly volatile GARCH processes. Interestingly, it was seen also that the use of consistent-threshold estimation increases the degree of size distortion exhibited. However, it was the examination of the symmetry hypothesis which provided the more dramatic and important results. The most crucial result concerned the substantial oversizing of the symmetry test for the TAR and MTAR models irrespective of both the covariance matrix estimator employed and the presence of GARCH. This finding indicated that a spurious inference of asymmetric adjustment could be noted with a high degree of probability when applying these tests and strongly rejected the use of the F-distribution for this test as routinely employed by practitioners (see [17, 20, 21]). Under the use of a deterministic threshold it was found that the symmetry test under the TAR model was undersized in the absence of GARCH, while the test under the MTAR model had approximately correct size. In the presence of GARCH behaviour, the most noticeable finding concerned the subsequent oversizing of the symmetry test for the MTAR model, although this was noted for the model using the standard OLS, rather than corrected, covariance matrix estimator.

In summary, the results of the present paper indicate that the routine application of TAR and MTAR unit root tests to economic and financial time series can result in incorrect or spurious inferences. While the probability of a spurious inference concerning the unit root hypothesis is most likely to occur when a corrected covariance matrix estimator is employed in the examination of time series processes exhibiting GARCH behaviour, the symmetry test has been shown to exhibit spurious rejection under consistent-threshold estimation even in the absence of GARCH. The results obtained indicate that practitioners should exercise care when applying TAR and MTAR unit root tests to economic and financial time series, particularly when considering the secondary test of symmetry where use of the F-distribution is clearly inappropriate despite its routine use in empirical analyses. As result of the above findings, an obvious future avenue of research concerns the development of asymmetric unit root tests which capture and draw upon any GARCH behaviour exhibited by data. This work, which can be viewed as a development of the studies of Seo [5] and Cook [6], is the subject of ongoing research.

REFERENCES

1. R. F. Engle, "Autoregressive conditional heteroscedasticity with estimates of the variance of United Kingdom inflation," Econometrica, vol. 50, no. 4, pp. 987–1007, 1982
2. T. Bollerslev, "Generalized autoregressive conditional heteroskedasticity," Journal of Econometrics, vol. 31, no. 3, pp. 307–327, 1986

3. S. Taylor, Modelling Financial Time Series, John Wiley & Sons, New York, NY, USA, 1986.

4. T. G. Andersen and T. Bollerslev, "Answering the skeptics: yes, standard volatility models do provide accurate forecasts," International Economic Review, vol. 39, no. 4, pp. 885–905, 1998

5. B. Seo, "Distribution theory for unit root tests with conditional heteroskedasticity," Journal of Econometrics, vol. 91, no. 1, pp. 113–144, 1999.

6. S. Cook, "Joint maximum likelihood estimation of unit root testing equations and GARCH processes: some finite-sample issues," Mathematics and Computers in Simulation, vol. 77, no. 1, pp. 109–116, 2008

7. S. Cook, "A re-examination of the stationarity of inflation," Journal of Applied Econometrics, vol. 24, no. 6, pp. 1047–1053, 2009.

8. K. Kim and P. Schmidt, "Unit root tests with conditional heteroskedasticity," Journal of Econometrics, vol. 59, no. 3, pp. 287–300, 1993.

9. S. Cook, "The impact of GARCH on asymmetric unit root tests," Physica A, vol. 369, no. 2, pp. 745–752, 2006.

10. S. Cook, "The robustness of modified unit root tests in the presence of GARCH," Quantitative Finance, vol. 6, no. 4, pp. 359–363, 2006.

11. C. Granger and D. Orr, "Infinite variance and research strategy in time series analysis?" Journal of the American Statistical Association, vol. 67, pp. 275–285, 1972.

12. M. Loretan and P. C. B. Phillips, "Testing the covariance stationarity of heavy-tailed time series: an overview of the theory with applications to several financial datasets," Journal of Empirical Finance, vol. 1, no. 2, pp. 211–248, 1994

13. R. S. Deo, "On estimation and testing goodness of fit for m-dependent stable sequences," Journal of Econometrics, vol. 99, no. 2, pp. 349–372, 2000.

14. S. I. Resnick, Heavy-Tail Phenomena, Springer Series in Operations Research and Financial Engineering, Springer, New York, NY, USA, 2007.

15. S. T. Rachev, S. Mittnik, and J.-R. Kim, "Time series with unit roots and infinite-variance disturbances,"Applied Mathematics Letters, vol. 11, no. 5, pp. 69–74, 1998.

16. K. D. Patterson and S. M. Heravi, "The impact of fat-tailed distributions on some leading unit roots tests," Journal of Applied Statistics, vol. 30, no. 6, pp. 635–667, 2003.

17. W. Enders and C. W. J. Granger, "Unit-root tests and asymmetric adjustment with an example using the term structure of interest rates," Journal of Business and Economic Statistics, vol. 16, no. 3, pp. 304–311, 1998.

18. H. White, "A heteroskedasticity-consistent covariance matrix estimator and a direct test for heteroskedasticity," Econometrica, vol. 48, no. 4, pp. 817–838, 1980

19. W. K. Newey and K. D. West, "Automatic lag selection in covariance matrix estimation," Review of Economic Studies, vol. 61, no. 4, pp. 631–653, 1994.

20. J. E. Payne and H. Mohammadi, "Are adjustments in the U.S. budget deficit asymmetric? Another look at sustainability," Atlantic Economic Journal, vol. 34,

no. 1, pp. 15–22, 2006.

21. M. A. Thompson, "Asymmetric adjustment in the prime lending-deposit rate spread," Review of Financial Economics, vol. 15, no. 4, pp. 323–329, 2006.

22. D. A. Dickey and W. A. Fuller, "Distribution of the estimators for autoregressive time series with a unit root," Journal of the American Statistical Association, vol. 74, no. 366, part 1, pp. 427–431, 1979.

23. H. Tong, Threshold Models in Nonlinear Time Series Analysis, vol. 21 of Lecture Notes in Statistics, Springer, New York, NY, USA, 1983.

24. H. Tong, Non-Linear Time-Series: A Dynamical Approach, Oxford University Press, Oxford, UK, 1990.

25. W. Enders, "Improved critical values for the Enders-Granger unit-root test," Applied Economics Letters, vol. 8, no. 4, pp. 257–261, 2001.

26. K. S. Chan, "Consistency and limiting distribution of the least squares estimator of a threshold autoregressive model," The Annals of Statistics, vol. 21, no. 1, pp. 520–533, 1993.

27. S. Cook and N. Manning, "The power of asymmetric unit root tests under threshold and consistent-threshold estimation," Applied Economics, vol. 35, no. 14, pp. 1543–1550, 2003

28. F. C. Drost and T. E. Nijman, "Temporal aggregation of GARCH processes," Econometrica, vol. 61, no. 4, pp. 909–927, 1993

29. R. Engle and A. Patton, "What makes a good volatility model?" Quantitative Finance, vol. 1, pp. 237–245, 2001.

CITATION

Steve Cook, "Simulation Analysis of Threshold Autoregressive Unit Root Tests," ISRN Probability and Statistics, vol. 2012, Article ID 649134, 12 pages, 2012. doi:10.5402/2012/649134.

Neural Networks Models for Stock Price Prediction

Ayodele Ariyo Adebiyi[1], Aderemi Oluyinka Adewumi[1], and Charles Korede Ayo[2]
[1]School of Mathematics, Statistics & Computer Science, University of KwaZulu-Natal, Westville, Durban, South Africa
[2]Department of Computer and Information Sciences, Covenant University, Ota, Nigeria

ABSTRACT

This paper examines the forecasting performance of ARIMA and artificial neural networks model with published stock data obtained from New York Stock Exchange. The empirical results obtained reveal the superiority of neural networks model over ARIMA model. The findings further resolve and clarify contradictory opinions reported in literature over the superiority of neural networks and ARIMA model and vice versa.

INTRODUCTION

Several research studies on stock predictions have been conducted with various solution techniques proposed over the years. The prominent techniques fall into two broad categories, namely, statistical and soft computing techniques. Statistical techniques include, among others, exponential smoothing, autoregressive integrated moving average (ARIMA), and generalized autoregressive conditional heteroskedasticity (GARCH) volatility [1]. The ARIMA model, also known as the Box-Jenkins model or methodology, is commonly used in analysis and forecasting. It is widely regarded as the most efficient forecasting technique in social science and is used extensively for time series. The use of ARIMA for forecasting time series is essential with uncertainty as it does not assume knowledge of any underlying model or relationships as in some other methods. ARIMA essentially relies on past values of the series as well as previous error terms for forecasting [2, 3]. However, ARIMA models are relatively more robust and efficient than more complex structural models in relation to short-run forecasting [3].

Artificial neural networks (ANNs) as a soft computing technique are the most accurate and widely used as forecasting models in many areas including social, engineering, economic, business, finance, foreign exchange, and stock problems [4–8]. Its wide usage is due to the several distinguishing features of ANNs that make them attractive to both researchers and industrial practitioners. As stated in [4], ANNs are data-driven, self-adaptive methods with few prior assumptions. They are also good predictor with the ability to make generalized observations from the results learnt from original data, thereby permitting correct inference of the latent part of the population. Furthermore, ANNs are universal approximator as a network can efficiently approximate a continuous function to the desired level of accuracy. Finally, ANNs have been found to be very efficient in solving nonlinear problems including those in real world [4]. This is in contrast to many traditional techniques for time series predictions, such as ARIMA, which assume that the series are generated from linear processes and as a result might be inappropriate for most real-world problems that are nonlinear [5, 6]. There is growing need to solve highly nonlinear, time-variant problems as many applications such as stock markets are nonlinear with uncertain behaviour that changes with time [7, 8]. ANNs are known to provide competitive results to various traditional time series models such as ARIMA model [4, 9–11]. In this paper, the performance of ANN and ARIMA models is studied and compared for a case of stock prediction, which also further clarify and/or confirm contradictory opinions reported in literature about superiority of each of the model over one another.

The rest of the paper is organized as follows: Section 2 presents some related works on the comparison of the ARIMA and ANNs model, while the methodology used in this work is presented in Section 3.

Section 4presents and discusses the experimental results obtained in this work, while useful conclusions are provided in Section 5.

RELATED WORKS

The search for efficient stock price prediction techniques is profound in literature. This is motivated partly by the dynamic nature of the problem as well as the need for better results. Tansel et al. [12] compared the performance of linear optimization, ANNs, and genetic algorithms (GAs) in modelling time series data based on modelling accuracy, convenience, and computational time. The study revealed that linear optimization techniques gave the best estimates with GAs providing similar results if the boundaries of the parameters and the resolution were carefully selected, while NNs gave the worst estimates. The work reported in [13] also compared the forecasting performance of ARIMA and ANN models in forecasting Korean Stock Price Index. The ARIMA model generally provided more accurate forecasts than the back-propagation neural network (BPNN) model used. This is more pronounced for the midrange forecasting horizons. Merh et al. [14] presented a comparison between hybrid approaches of ANN and ARIMA for Indian stock trend forecasting with many instances of the ARIMA predicted values shown to be better than those of the ANNs predicted values in relation to the actual stock value. Sterba and Hilovska [15] argued that ARIMA model and ANN model achieved good prediction performance in many real-world applications especially time series prediction. Experimental results obtained by the authors further revealed that ARIMA model generally performs better in the prediction of linear time series, while ANNs perform better in the prediction of nonlinear time series. In a similar study for financial forecasting reported in [16], ANNs model was shown to perform better than ARIMA model in value forecasting, while ARIMA model performed better than ANNs in directional forecasting.

Yao et al. [17] compared the stock forecasting performance of ANN and ARIMA models and showed that the ANN model obtained better returns than the conventional ARIMA models Similarly, Hansen et al.

[18] compared the prediction performance of ANNs and ARIMA on time series prediction to show that the ANNs outperformed ARIMA in predicting stock movement direction as the latter was able to detect hidden patterns in the data used. Prybutok et al. [19] also compared the forecasting performance of ANN and ARIMA model in forecasting daily maximum ozone concentration. Empirical results obtained also showed that the ANN model is superior to the ARIMA model. Wijaya et al. [20] did similar comparison based on the Indonesia stock exchange and got better accuracy with ANN than the ARIMA model. More literature has shown the prevalent use of ANNs as an effective tool for stock price prediction [10, 21–29]. This makes ANN a promising technique or potential hybrid for the prediction of movement in time series.

However, literature has shown different view on the relative performance and superiority of ARIMA and ANNs models to time series prediction, especially for different data used; hence the need for further study that can help unified a coherent view on the better methodology. This paper therefore seeks to further clarify contradictory opinions reported in literature on the superiority of ANN model over ARIMA model and vice versa in the effective prediction of stock prices. Results obtained are based on empirical study on time series stock prediction using data from the New York Stock Exchange (NYSE).

METHODOLOGY

The research methodology used in this study is summarized below. The study used published stock data from NYSE on ARIMA and ANN models developed. EViews software and Matlab Neural Network Tools Box version 7 were used for ARIMA and ANNs models, respectively.

Input Data

The data used in this research work were historical daily stock prices. The stock data consists of open price, low price, high price, close price, and volume traded. The open price is the opening price of the index (PoI) at the start of the trading day, the low price represents the minimum PoI during the trading day, the high price represents the maximum PoI during the trading day, and the closing price indicates the PoI when the market closes. In this research the closing price is chosen to represent the PoI to be modeled and predicted. This is because the closing price reflects all the activities of the index of the day.

ARIMA (p,d,q) Model Development for Stock Price of Dell Incorporation

This study used the Dell Inc. stock data used that covered the period from August 17, 1988, to February 25, 2011, having a total number of 5680 observations. It was observed that the original pattern of the time series of the index is not stationary. The time series have random walk pattern and vary randomly with no global trend or seasonality pattern observed.

A correlogram is used to determine whether a particular series is stationary or nonstationary. Usually, a stationary time series will give an autocorrelation function (ACF) that decay rapidly from its initial value of unity at zero lag. In the case of nonstationary time series, the ACF dies out gradually over time. The correlogram of the time series of Dell stock index was observed to be nonstationary as the ACF dies down extremely slowly. Differencing is used to make this nonstationary time series become stationary. The value of difference (d) is determined by the number of times the differencing is performed on the time series.

In order to construct the best ARIMA model for Dell stock index, the autoregressive (p) and moving average (q) parameters have to be effectively determined for an effective model. To determine the best model, we set the criteria as follows (also depicted in Table 1): relatively small Bayesian Information Criterion (BIC) and Standard Error of regression (SER), relatively high adjusted R^2. The Q-statistics and correlogram done showed no significant pattern left in the ACFs and partial autocorrelation functions (PACFs) of the residuals which implies that the residual of the selected model is white noise.

Table 1: ARIMA (1, 0, 0) estimation output with CLOSE of Dell index

Dependent variable: CLOSE
Method: least squares
Date: 03/21/11 Time: 15:54
Sample (adjusted): 8/18/1988–2/25/2011
Included observations: 5679 after adjustments
Convergence achieved after 4 iterations

Variable	Coefficient	Standard error	t-statistic	Prob.
C	34.11484	6.028238	5.659173	0.0000
AR (1)	0.994802	0.001346	739.1456	0.0000
R-squared	0.989716	Mean dependent variable		33.91262
Adjusted R-squared	0.989714	S.D. dependent variable		23.28046
S.E. of regression	2.361101	Akaike info criterion		4.556485
Sum squared residual	31648.13	Schwarz criterion		4.558825
Log likelihood	−12936.14	F-statistic		546336.2
Durbin-Watson static	2.015870	Prob. (F-statistic)		0.000000
Inverted AR roots	0.99			

Table 2 shows the different parameters p and q in the ARIMA model. ARIMA (1,0,0) is considered the best for Dell stock index as shown in Table 1.

Table 2: Statistical results of different ARIMA parameters for Dell stock index

ARIMA	BIC	Adjusted R^2	SER
(1, 0, 0)	4.5588	0.9897	2.3611
(1, 0, 1)	4.5602	0.9897	2.3612
(2, 0, 0)	5.2389	0.9796	3.3174
(0, 0, 1)	7.8883	0.7127	12.4770
(0, 0, 2)	7.9369	0.6984	12.7839
(1, 1, 0)	4.5615	−0.0000	2.3642
(0, 1, 0)	4.5599	0.0000	2.3639
(0, 1, 1)	4.5615	−0.0000	2.3642
(1, 1, 2)	4.5630	−0.0002	2.3644
(2, 1, 0)	4.5617	−0.0001	2.3645
(2, 1, 2)	4.5610	0.0019	2.3621

In forecasting form, the best model selected can be expressed as follows:

$$Y_t = \phi_1 Y_{t-1} + \theta_0 + \varepsilon_t, \tag{1}$$

where $\varepsilon_t = Y_t - \hat{Y}_t$ is the difference between the actual value and the forecast value of the series.

ANN Model Construction for the Dell Stock Index

This study employed a three-layer (one hidden layer) multilayer perceptron model trained with back-propagation algorithm. The ANN model used for the nonlinear data is represented as follows:

$$y_t = w_0 + \sum_{j=1}^{q} w_j \cdot g\left(w_{0j} + \sum_{i=1}^{p} w_{ij} \cdot y_{t-1}\right) + \varepsilon_t,$$

(2)

where $(i = 0,1,2,....,p, j = 1,2,......,q)$ and $w_j (j = 0,1,2,....,q)$ are the connection weights, p is the number of input nodes, and q is the number hidden nodes. Ten input variables, each grouped into two as inputs for day $i-1$ and day $i-2$ were supplied into the model. These variables are the opening price (o_{i-1}, o_{i-2}), daily high price (H_{i-1}, H_{i-2}), daily low price (L_{i-1}, L_{i-2}), daily closing price (C_{i-1}, C_{i-2}), and trading volume (V_{i-1}, V_{i-2}).

The creation of the ANN predictive model with Matlab for the Dell stock index involves the following.

i. Creating the Network Topology: This involves the selection of the number of input neurons (in this case 10 inputs), the number of hidden layers, the number of hidden neurons in the hidden layer (see Table 3), and the number of output neurons (one, in this case).

ii. Training the Network: This involves selecting the network type/ training algorithm, in our case feed-forward back-propagation algorithm, inputting the training and target data, selecting the training function (TRAINGDM), selecting the adaptation learning function (LEARNGDM), selecting the performance function (MSE), and selecting the transfer function (TANSIG).

The training parameters were set as follows: learning rate = 0.01, momentum term = 0.9, and epoch size = 1000, 2000, 5000. Finally, the network was tested with the data set to estimate its generalization ability.

Table 3: Statistical performance of ANN model of Dell stock index*

Network structure	MSE		
	1000 epochs	2000 epochs	5000 epochs
10-10-1	0.129054	0.112363	0.093539
10-11-1	0.144086	0.108245	0.090521
10-12-1	0.125668	0.099301	0.088157
10-13-1	0.148646	0.115732	0.092649
10-14-1	0.141474	0.099241	0.085206
10-15-1	0.118226	0.096651	0.083664
10-16-1	0.116773	0.099222	0.080534
10-17-1	0.097826	0.085111	0.071589
10-18-1	0.119719	0.093576	0.079150

*The bold characters indicate the best results for each of the epoch sessions.

To determine the best performing model, simulation experiment was run on different ANN model configurations. Both training and testing data were carefully selected. However, the training was not done with test data. The model was trained with 1000, 2000, and 5000 epochs, respectively, while the mean squared error (MSE) for each training session of the different network structure was noted.

Figure 1 is the graph of network training showing the best performance in each of the network structure models in the different training sessions. The network structure that returns the smallest MSE in each of the models was adjudged the best model that can give the best accurate prediction. Similarly, Table 3 presents the outcome of the various training sessions in each of the ANN network structure. It was observed in most cases that the best model was obtained when the network was well trained.

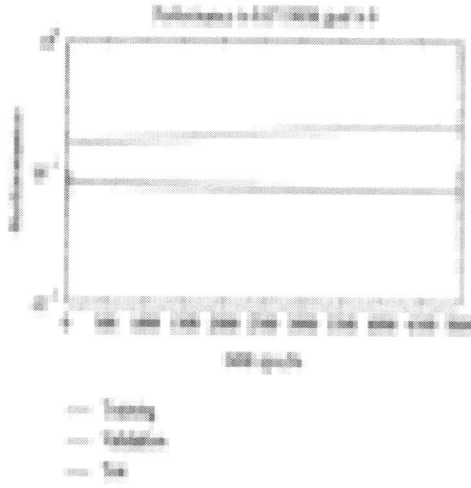

Figure 1: Graph of the best result achieved in network training of ANN of Dell index.

EXPERIMENTAL RESULTS AND DISCUSSION

The tools for simulation of the models are Matlab 2007 and EViews software for ANN model and ARIMA model, respectively. The results obtained are presented in the subsection below.

Result of ARIMA Model

We experimented with different parameters of autoregressive (p) and moving average (q) in order to determine the best model that will give best forecast as indicated in Table 2. ARIMA (1,0,0) is considered the best for Dell stock index as shown in Table 1; hence it was selected as the best model based on the criteria listed in the previous section. The actual stock price and predicted values are presented in Table 4, while Figure 2 gives the graph of predicted price against actual stock price to see the performance of the ARIMA model selected. From the predicted values, it was observed that a constant number is added to the subsequent values from the previous value and this accounted for the linear graph of the predicted values in Figure 2. However, the forecast error is quite low and impressive as the predicted values are close to the actual values and move in the direction of the forecast values in many instances as shown in Figure 2, which depicts the correlation of the level of accuracy. The forecast error is determined by

Table 4: Sample of empirical results of ARIMA (1, 0, 0) of Dell stock index

Sample period	Actual values	Predicted values	Forecast error
01/03/2010	13.57	13.35	0.016212
02/03/2010	13.68	13.46	0.016082
03/03/2010	13.71	13.56	0.010941
04/03/2010	13.67	13.67	0
05/03/2010	13.88	13.78	0.007205
08/03/2010	14.01	13.88	0.009279
09/03/2010	14.18	13.99	0.013399
10/03/2010	14.31	14.09	0.015374
11/03/2010	14.21	14.2	0.000704
12/03/2010	14.26	14.3	−0.00281
15/03/2010	14.26	14.4	−0.00982
16/03/2010	14.3	14.51	−0.01469
17/03/2010	14.59	14.61	−0.00137
18/03/2010	14.55	14.71	−0.011
19/03/2010	14.41	14.81	−0.02776
22/03/2010	14.62	14.91	−0.01984
23/03/2010	15.22	15.01	0.013798
24/03/2010	14.99	15.11	−0.00801
25/03/2010	14.87	15.21	−0.02286

Sample period	Actual values	Predicted values	Forecast error
26/03/2010	14.99	15.31	−0.02135
29/03/2010	14.96	15.4	−0.02941
30/03/2010	14.97	15.5	−0.0354
31/03/2010	15.02	15.6	−0.03862

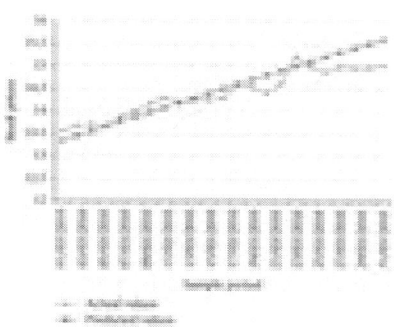

Figure 2: Graph of actual stock price versus predicted values for Dell stock index using ARIMA.

Results of ANN Model

After several experiments with different network architectures based on our ANN algorithm, the network structure that returns the smallest MSE was noted to give the best forecasting accuracy with the test data. The MSE recorded in the experiments are presented in Table 3, from where we observed that 10-17-1 (10 input neurons, 17 hidden neurons, and 1 output neuron) is the predictive model with the most accurate daily price prediction. The results presented in Table 5 were the findings from testing period (out of sample test data), while Figure 3 illustrates the correlation of the level accuracy. The forecast error of ANN model is equally low which demonstrated good forecast performance as indicated in Table 5.

Table 5: Sample results of ANN model for Dell stock index

Sample period	Actual value	Predicted value	Forecast error
01/03/2010	13.57	13.16	0.03021
02/03/2010	13.68	13.55	0.0095
03/03/2010	13.71	13.7	0.00073
04/03/2010	13.67	13.55	0.00878
05/03/2010	13.88	13.53	0.02522
08/03/2010	14.01	13.89	0.00857
09/03/2010	14.18	13.92	0.01834
10/03/2010	14.31	13.85	0.03215
11/03/2010	14.21	14.18	0.00211
12/03/2010	14.26	14.31	−0.0035
15/03/2010	14.26	14.15	0.00771
16/03/2010	14.3	14.49	−0.0133
17/03/2010	14.59	14.5	0.00617
18/03/2010	14.55	14.25	0.02062
19/03/2010	14.41	14.28	0.00902
22/03/2010	14.62	14.67	−0.0034
23/03/2010	15.22	15.19	0.00197
24/03/2010	14.99	14.66	0.02201
25/03/2010	14.87	14.96	−0.0061
26/03/2010	14.99	14.75	0.01601
29/03/2010	14.96	14.89	0.00468
30/03/2010	14.97	15.01	−0.0027
31/03/2010	15.02	14.97	0.00333

Comparison of ARIMA and ANN Model

From the empirical results presented in Table 6 and Figure 4, we observed that the forecasting accuracy level of the ANN model compared with that of the ARIMA model is not quite significant. It can be

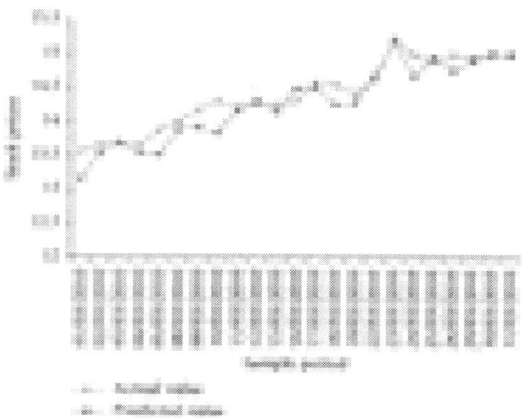

Figure 3: Graph of ANN model of predicted values against actual values for Dell stock index.

argued that both models achieved good forecast performance judging from the forecast error of both models which are quite low. This finding agrees with the work of [15]. However, the performance of ANN model is better than ARIMA model in terms of forecasting accuracy on many occasions from the test data. Results of Figure 4show that the ANN model is better than the ARIMA model for stock price prediction. We also observed that the pattern of ARIMA model is directional, which accounted for the linear pattern observed in the graph of Figure 2, while ANN model is toward value forecasting. This finding also agrees with the work of [16]. Statistical test was carried out, which also showed that there is no significant difference between the actual and predicted values of the two models as the p values of ANN and ARIMA are 0.439 and 0.604, respectively. Notwithstanding, ANN is still better. Hence, this research work also further clarifies the contrary opinions reported in literature about the superiority of ANN model over ARIMA model in time series prediction.

CONCLUSIONS

The empirical results obtained with published stock data on the performance of ARIMA and ANN model to stock price prediction have been presented in this study. The performance of the ANN predictive

Table 6: Sample results of ANN and ARIMA models for Dell stock index

Sample period	Actual value	Predicted values		Forecast error	
		ANN	ARIMA	ANN	ARIMA
01/03/2010	13.57	13.16	13.35	0.03021	0.016212
02/03/2010	13.68	13.55	13.46	0.0095	0.016082
03/03/2010	13.71	13.7	13.56	0.00073	0.010941
04/03/2010	13.67	13.55	13.67	0.00878	0
05/03/2010	13.88	13.53	13.78	0.02522	0.007205
08/03/2010	14.01	13.89	13.88	0.00857	0.009279
09/03/2010	14.18	13.92	13.99	0.01834	0.013399
10/03/2010	14.31	13.85	14.09	0.03215	0.015374
11/03/2010	14.21	14.18	14.2	0.00211	0.000704
12/03/2010	14.26	14.31	14.3	−0.0035	−0.00281
15/03/2010	14.26	14.15	14.4	0.00771	−0.00982
16/03/2010	14.3	14.49	14.51	−0.0133	−0.01469
17/03/2010	14.59	14.5	14.61	0.00617	−0.00137
18/03/2010	14.55	14.25	14.71	0.02062	−0.011
19/03/2010	14.41	14.28	14.81	0.00902	−0.02776
22/03/2010	14.62	14.67	14.91	−0.0034	−0.01984
23/03/2010	15.22	15.19	15.01	0.00197	0.013798
24/03/2010	14.99	14.66	15.11	0.02201	−0.00801
25/03/2010	14.87	14.96	15.21	−0.0061	−0.02286
26/03/2010	14.99	14.75	15.31	0.01601	−0.02135
29/03/2010	14.96	14.89	15.4	0.00468	−0.02941
30/03/2010	14.97	15.01	15.5	−0.0027	−0.0354
31/03/2010	15.02	14.97	15.6	0.00333	−0.03862

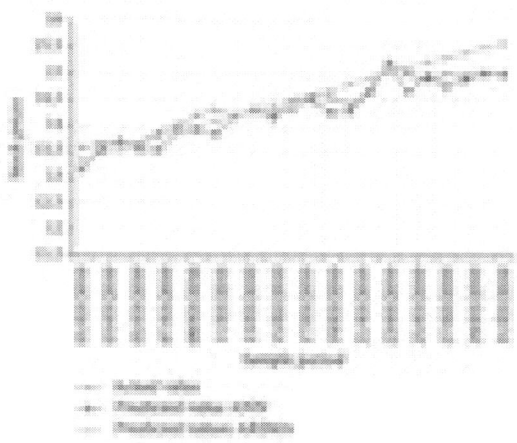

Figure 4: Graph of predicted values of ARIMA and ANN model against actual stock price.

model developed in this study was compared with the conventional Box-Jenkins ARIMA model, which has been widely used for time series forecasting. Our findings revealed that both ARIMA model and ANN model can achieve good forecast in application to real-life problems and thus can be effectively engaged profitably for stock price prediction. We also observed that the pattern of ARIMA forecasting models is directional. The developed stock price predictive model with the ANN-based approach demonstrated superior performance over the ARIMA models; indeed, the actual and predicted values of the developed stock price predictive model are quite close. In future studies, hybrid of intelligent techniques similar to that reported in [11, 15,30] can be engaged to improve existing predictive models with recent stock data and more stock index.

REFERENCES

1. J. J. Wang, J. Z. Wang, Z. G. Zhang, and S. P. Guo, "Stock index forecasting based on a hybrid model,"Omega, vol. 40, no. 6, pp. 758–766, 2012.
2. B. G. Tabachnick and L. S. Fidell, Using Multivariate Statistics, Pearson Education, Upper Saddle River, NJ, USA, 4th edition, 2001.

3. A. Meyler, G. Kenny, and T. Quinn, "Forecasting Irish Inflation Using ARIMA Models," Technical Paper 3/RT/1998, Central Bank of Ireland Research Department, 1998.

4. M. Khashei and M. Bijari, "An artificial neural network (p, d, q) model for time-series forecasting,"Expert Systems with Applications, vol. 37, no. 1, pp. 479–489, 2010. ·

5. G. Zhang, B. Patuwo, and M. Y. Hu, "Forecasting with artificial neural networks: the state of the art,"International Journal of Forecasting, vol. 14, no. 1, pp. 35–62, 1998.

6. M. Khashei, M. Bijari, and G. A. R. Ardali, "Improvement of auto-regressive integrated moving average models using fuzzy logic and artificial neural networks (ANNs)," Neurocomputing, vol. 72, no. 4–6, pp. 956–967, 2009. ·

7. R. Fuller, Neural Fuzzy System, Abo Akademic University, 1995.

8. E. Khan, "Neural fuzzy based intelligent systems and applications," in Fusion of Neural Networks, Fuzzy Systems, and Genetic Algorithms Industrial Application, C. J. Lakhmi and N. M. Martin, Eds., The CRC Press International Series on Computational Intelligence, pp. 107–139, CRC Press, New York, NY, USA, 2000.

9. Y. Chen, B. Yang, J. Dong, and A. Abraham, "Time-series forecasting using flexible neural tree model," Information Sciences, vol. 174, no. 3-4, pp. 219–235, 2005. ·

10. F. Giordano, M. La Rocca, and C. Perna, "Forecasting nonlinear time series with neural network sieve bootstrap," Computational Statistics and Data Analysis, vol. 51, no. 8, pp. 3871–3884, 2007.

11. A. Jain and A. M. Kumar, "Hybrid neural network models for hydrologic time series forecasting,"Applied Soft Computing Journal, vol. 7, no. 2, pp. 585–592, 2007.

12. I. N. Tansel, S. Y. Yang, G. Venkataraman, A. Sasirathsiri, W. Y. Bao, and N. Mahendrakar, "Modeling time series data by using neural networks and genetic algorithms," in Smart Engineering System Design: Neural Networks, Fuzzy Logic, Evolutionary Programming, Data Mining, and Complex Systems: Proceedings of the Intelligent Engineering Systems Through Artificial Neural Networks, C. H. Dagli, A. L. Buczak, J. Ghosh, M. J. Embrechts, and O. Erosy, Eds., vol. 9, pp. 1055–1060, ASME Press, New York, NY, USA, 1999.

13. C. K. Lee, Y. Sehwan, and J. Jongdae, "Neural network model versus SARIMA model in forecasting Korean stock price index (KOSPI)," Issues in Information System, vol. 8, no. 2, pp. 372–378, 2007.

14. N. Merh, V. P. Saxena, and K. R. Pardasani, "A comparison between hybrid approaches of ANN and ARIMA for Indian stock trend forecasting," Journal of Business Intelligence, vol. 3, no. 2, pp. 23–43, 2010.

15. J. Sterba and K. Hilovska, "The implementation of hybrid ARIMA neural network prediction model for aggregate water consumption prediction," Aplimat—Journal of Applied Mathematics, vol. 3, no. 3, pp. 123–131, 2010.

16. A. G. Lahane, "Financial forecasting: comparison of ARIMA, FFNN and SVR,"

2008,http://www.it.iitb.ac.in/~ashishl/files/MTechProjectPresentation.pdf.

17. J. T. Yao, C. L. Tan, and H. L. Poh, "Neural networks for technical analysis: a study on KLCI,"International Journal of Theoretical and Applied Finance, vol. 2, no. 2, pp. 221–241, 1999.

18. J. V. Hansen, J. B. Mcdonald, and R. D. Nelson, "Time series prediction with genetic-algorithm designed neural networks: an empirical comparison with modern statistical models," Computational Intelligence, vol. 15, no. 3, pp. 171–184, 1999.

19. V. R. Prybutok, J. Yi, and D. Mitchell, "Comparison of neural network models with ARIMA and regression models for prediction of Houston›s daily maximum ozone concentrations," European Journal of Operational Research, vol. 122, no. 1, pp. 31–40, 2000.

20. Y. B. Wijaya, S. Kom, and T. A. Napitupulu, "Stock price prediction: Comparison of Arima and artificial neural network methods—an Indonesia stock›s case," in Proceedings of the 2nd International Conference on Advances in Computing, Control and Telecommunication Technologies (ACT ‹10), pp. 176–179, Jakarta, Indonesia, December 2010.

21. P. M. Tsanga, P. Kwoka, S. O. Choya et al., "Design and implementation of NN5 for Hong stock price forecasting," Journal of Engineering Applications of Artificial Intelligence, vol. 20, no. 4, pp. 453–461, 2007.

22. T. H. Roh, "Forecasting the volatility of stock price index," Expert Systems with Applications, vol. 33, no. 4, pp. 916–922, 2007.

23. H. Al-Qaheri, A. E. Hassanien, and A. Abraham, "Discovering stock price prediction rules using rough sets," Neural Network World, vol. 18, no. 3, pp. 181–198, 2008.

24. B. Vanstone and G. Finnie, "An empirical methodology for developing stockmarket trading systems using artificial neural networks," Expert Systems with Applications, vol. 36, no. 3, pp. 6668–6680, 2009.

25. S. K. Mitra, "Optimal combination of trading rules using neural networks," International Business Research, vol. 2, no. 1, pp. 86–99, 2009.

26. G. S. Atsalakis and K. P. Valavanis, "Forecasting stock market short-term trends using a neuro-fuzzy based methodology," Expert Systems with Applications, vol. 36, no. 7, pp. 10696–10707, 2009.

27. M. M. Mostafa, "Forecasting stock exchange movements using neural networks: empirical evidence from Kuwait," Expert Systems with Applications, vol. 37, no. 9, pp. 6302–6309, 2010.

28. E. Hadavandi, H. Shavandi, and A. Ghanbari, "Integration of genetic fuzzy systems and artificial neural networks for stock price forecasting," Knowledge-Based Systems, vol. 23, no. 8, pp. 800–808, 2010.

29. T. H. Yu and K. H. Huarng, "A neural network-based fuzzy time series model to improve forecasting," Expert Systems with Applications, vol. 37, no. 4, pp. 3366–3372, 2010.

30. A. O. Adewumi and A. Moodley, "Comparative results of heuristics for portfolio

selection problem," in Proceedings of the IEEE Conference on Computational Intelligence for Financial Engineering & Economics (CIFEr ‹12), pp. 1–6, New York, NY, USA, March 2012.

CITATION

Ayodele Ariyo Adebiyi, Aderemi Oluyinka Adewumi, and Charles Korede Ayo, "Comparison of ARIMA and Artificial Neural Networks Models for Stock Price Prediction," Journal of Applied Mathematics, vol. 2014, Article ID 614342, 7 pages, 2014. doi:10.1155/2014/614342.

Wavelet Transform in Forecasting Banking Sector

[a]S. Al Wadi, [b]Abdulkareem Hamarsheh and [c]Hazem Alwadi

[a,b]Department of mathematics, Faculty of Science Al Hussien bin Talal University, Ma'an, Jordan
[c]School of Social Science Jordan University of Science and Technology, Irbid, Jordan

ABSTRACT

In this paper, we present the advantages of Maximum overlapping Discrete Wavelet Transform (MODWT) in improving the forecasting accuracy financial time series data. Amman stock market (ASE) in Jordan was selected as a tool to show the ability of MODWT in forecasting financial time series, using Banking sector. Experimentally, this article suggests a novel technique for forecasting the banking data based on MODWT and ARIMA model. Daily return data from 1993 until 2009 is used for this study.

INTRODUCTION

During the last three decades the banking sector was very stable in Jordan, very few number of banks have exited from the market. Since this sector has a very nice environment for Growing, well capitalized, liquid and profitable and privately owned, Open to external investors. Moreover, this sector has Comprehensive banking services: retail, corporate, Islamic, and e-banking and loans, Payment System: Real Time Gross Settlements and Electronic Cheque Clearing System. Generally, this sector has regional banks: 137 branches outside Jordan.

These reasons motivate the researches to focus in analyzing the banking sector in Jordan as in [18], and improving the forecasting accuracy using MODWT as in this article. In recent years, stock markets forecasting is required for the investors and it has got very high attention in financial time series and financial researchers. The accurate forecasting of financial prices is an important issue in investment decision making.

However, financial time series data appears noisy and non-stationary [6,12]. The noise characteristic indicates the unavailability of complete information from past behavior of financial markets to fully capture the dependency between future and past prices. The information that is excluded in the forecasting model is considered as noise while the non-stationary characteristic indicates the distribution of financial time series changing over time. Therefore, financial time series forecasting is considered as one of the most challenging tasks of time series analysis.

There are many forecasting models that have been used in the forecasting literature, such as; simple moving average, linear regression, neural network, ARMA model and ARIMA model. In order to provide estimates for the future, these models analyze the historical data. Usually time series are not deterministic series. In this regard, MODWT will be used as an effective method to forecast the future events in banking sector in ASE, then selecting the best WT function in forecasting processes.

This paper is organized as follows. The next section describes the mathematical and literature review. Section 3 provides a description of data set. Section 4 describes the methodology. The experimental results are presented to demonstrate the effectiveness of MODWT in using banking data will be presented in section 5. In Section 6 we summarize our contributions and mention the conclusion

MATHEMATICAL AND LITERATURE REVIEW

Wavelet Transform

Wavelet analysis is a mathematical model that transforms the original signal (especially with time domain) into a different domain for analysis and processing [13,16]. This model is very suitable with the non-stationary data, i.e. mean and autocorrelation of the signal are not constant over time, that is well known, most of the financial time series data is non-stationary, that is why we applied MODWT.

In the MODWT and discrete wavelet transform (DWT) case, consider that the time domain is the original domain. Although, these models of transformation process from time domain to time scale domain, these processes are known as signal decomposition because a given signal is decomposed into several other signals with deferent levels of resolution. These processes allow recovering the original time domain signal without losing any information. MODWT has reverse process which is called the inverse MODW Tor signal reconstruction [4].

The MODWT and DWT is implemented using a multiresolution pyramidal decomposition technique. In fact, a recorded digitized time signal S(n) can be analyzed into its detailed cD1(n) and smoothed (approximations) cA1(n) signals using high-pass filter (HiF-D) and low-pass filter (LoF-D), respectively. Highpass filter has a band-pass response. Consequently, the filter signal cD1(n) is a detailed coefficient of S (n) and contains higher frequency components. While the approximation signal cA1(n) has a low-pass frequencies filter response. The decomposition of S (n) into cA1 (n) and cD1(n) is the first scale decomposition. Inversely, that is possible to perform the original signal from the approximations and details coefficients.

In this paper we will focus in the most famous types of discrete DWT which are HWT (Haar wavelet transform) and dWT (Daubechies wavelet transform), then compare them with MOWDT dynamic. The wavelets having compact support or narrow window function are suitable for local analysis of the signal. dWT, HWT and MODWT are compactly supported orthonormal [1].

Definition: [2,7] DWT can be defined by the following function:

$$\psi_{j,k}(t) = 2^{\frac{j}{2}} \psi(2^j t - k),$$

$$j, k \in Z; \ z = \{0, 1, 2, ...\}$$

where ψ is a real valued function having compactly supported, and

$$\int_{-\infty}^{\infty} \psi(t) dt = 0$$

Generally, the MODWT were evaluated by using dilation equations, given as:

$$\phi(t) = \sqrt{2} \sum_k l_k \phi(2t - k), \quad \psi(t) = \sqrt{2} \sum_k h_k \phi(2t - k)$$

Father and mother wavelets were defined by the last two equations where $\phi(2t-k)$ represents the father wavelet, and $\psi(t)$ represents the mother wavelet. Father wavelet defines the lower pass filter coefficients (h_k) and high pass filters coefficients (l_k) are defined as [3].

$$l_k = \sqrt{2} \int_{-\infty}^{\infty} \phi(t) \phi(2t - k) dt, \quad h_k = \sqrt{2} \int_{-\infty}^{\infty} \psi(t) \psi(2t - k) dt$$

HWT is the oldest and simplest example in the DWT and is defined as:

$$\psi^H(t) = \begin{cases} 1, & 0 \le t \le \dfrac{1}{2} \\ -1, & \dfrac{1}{2} \le t \le 1 \\ 0, & Otherwise \end{cases}$$

where $\psi_1(\omega)$ presents the HWT. HWT is the simplest and oldest DWT; it was improved by dWT in 1992. He developed the frequency – domain characteristics of the HMODWT. However, we do not have a specific formula for this method of MODWT. So, we tend to use the square gain function of their scaling filter, the square gain function was defined as [9].

$$g(f) = 2\cos^{l}(\pi f)\sum_{\iota=0}^{\frac{l}{2}-1}\left(\begin{array}{c}\frac{l}{2}-1+l\\ l\end{array}\right)\sin^{2l}(\pi f)$$

where l: Positive number and represents the length of the filter, for more details and examples about the MODWT mathematical model and its applications refer [2,8, 9].

As critically review. First, According to the past decade, few researchers has paying attention on the application of MODWT to solve financial issues such as study the banking data and forecasting in ASE. Second, for the last 10 years a number of comparative studies using different methodologies have been carried out using various MODWT functions alone as well as in combination with other models. However, no research conducted which is comparing between DWT and MODWT in contert of forecasting area using banking data. Therefore, the forecasting model will be used is ARIMA model, since this model is easy to combine with other models and can be used for long run of dataset.

ARIMA Model

ARMA is a suitable model for the stationary time series data, although most of the software uses least square estimation which requires stationary. To overcome this problem and to allow ARMA model to handle non-stationary data, the researchers investigate a special class for the non-stationary cata. This model is called Autoregressive Integrated Moving Average (ARIMA). This idea is to separate a nonstationary series one or more times until the time series becomes stationary, and then find the fit model. ARIMA model has got very high attention in the scientific world. This model is popularized by George Box and Gwilym Jenkins in 1970s [4]. There are a huge number of ARIMA models;

generally there are ARIMA (p, q, d) where: P: order of autoregressive part (AR), d: degree of first differentiation (I) and q: order of the first moving part (MA). Note that, if there is no differencing been done (d = 0), Then ARMA model can be got from ARIMA model [5]. The general mathematical ARIMA model can be defined as [11]:

Where:

t : Indexes time.

W_t : The response series Y_t or a difference of the response series.

μ: The mean term.

v : The backshift operator; that is, $vX_t = X_{t-1}$

ε (v) : The autoregressive operator.

β(v) : The moving-average operator

a_t : The independent disturbance, also called the random error.

DATA DESCRIPTION

In order to illustrate the effectiveness of MODWT and HWT and dWT the Amman Stock Market data sets are selected for discussion. We consider a daily return data for the time period from April 1993 (the days when stock exchanges were open) until December 2009 with a total of 4096 observations. In HWT and dWT the total number of observations for mathematical convenience is suggested to be divisible by 2^j. It means that the data should satisfy the condition of observations= 2^j, whereas this condition is not necessary for MODWT [7,17].

METHODOLOGY

We will present the criteria which have been used to make a fair comparison, and then the framework comparison will be presented with more details.

Prediction Accuracy Criteria

We have been adopted to compare the performance of the models within two types of accuracy criteria [14]:

- Root mean squared error (RMSE)

$$RMSE = \sqrt{\frac{\sum_{i=1}^{N}(\text{actual value} - \text{predicted value})^2}{N}}$$

- Mean absolute percentage error (MAPE).

$$MAE = \frac{1}{2}\sum_{i=1}^{N}\left|\frac{\text{actual value} - \text{predicted value}}{\text{actual value}}\right|.100\%$$

Where N represents the number of observations used for analysis.

Comparison framework:

The MODWT converts the data into two sets; approximation series (CA1 (n)) and details series (DA1 (n)). These two series present a better behavior. i.e. More stable in variance and no outliers than the original price series, then, they can be predicted more accurately. The reason for the better behavior of these two series is the filtering effect of the MODWT. In this paper the Approximation series has been used since this series behave as the main component of the transform. The procedure explained in this paper is as follows:

First, decompose through the MODWT, HWT and dWT the available historical return data. Second, Use a specific ARIMA model fitted to each one of the Approximation series to make the forecasting. Third, this technique is compared with an ARIMA model used directly to forecast the return data series by using the above criteria. Fourth, comparing the results for the entire model used in this paper

EXPERIMENTAL RESULTS

In this paper, the minimum value of MSE, RMSE and MAE is considered to select the best ARIMA model of the daily return data. All choices of ARIMA models for the return data are included in this test between (0, 0, 0) and (2, 2, 2). If we choose more than two, then there are more complicated conditions that should be satisfied. Also, if p and q are more than two, then Autocorrelation function (ACF) and partial Autocorrelation function (PACF) will be presented as an exponential decay. This means that ARIMA model becomes worthless and there is no importance.

Table1: Shows the statistical criteria for the ARIMA (p,d,q) model

Statistical fit	Value after transform via MODWT	Value after transform via HWT	Value after transform via dWT
RMSE	0.4	0.88	0.67
MAE	3.5%	10.44%	8.7%

The banking data for Amman stock market has been used as a case study. Price forecasting is performed using daily data. Moreover, for the sake of fair comparison the same sample data is selected. (From 1993-2009). The fit ARIMA model with HWT for the original return data is considered with RMSE equal to 0.88.as presented in Table 1, which is the worst model in this case study. While the fit ARIMA model for the transform data by using dWT is selected with RMSE equal to 0. 67 as presented in Table 1 also. Whereas, the forecasting is the best using MODWT since it has RMSE equal to 0.4. In order to further corroborate our findings and hence conclusions, this study also used another statistical criterion to analyze the small difference in RMSE, which is, MAPE. Table 1 shows the results using MAPE. When the results of these criteria (RMSE and MAPE) are taken collectively, the forecasting accuracy is improved using MODWT combined with a suitable ARIMA model compared to the forecasting accuracy using ARIMA model with HWT and dWT respectively.

CONCLUSIONS

This study implemented MODWT on ASE banking data. The success application of this study is in removal the outliers and irregular data. Therefore, in this empirical study the sample data set was experimentally tested in terms of forecasting accuracy and decomposition levels. The purpose of doing so was to find out, whether MODWT combined with a suitable ARIMA model produces more accurate forecasting compared to HWT and DWT with ARIMA model. And also to find out the event that occurred in banking sector in ASE during past 20 years. The findings using statistical prediction criteria RSME and MAPE have revealed that MODWT combined with ARIMA model is more accurate at forecasting. Therefore, this particular finding means that the forecasting can be improved using this modern model (MODWT).

REFERENCES

1. A.J. Rocha Reis, and A.P. Alves da Silva, Feature extraction via multiresolution analysis for short term load forecasting, IEEE Transactions on Power Systems, 20 (2005), 189–98.
2. Chang Chiann and Pedro A. Moretin, A Wavelet Analysis for Time Series, Non-parametric Statistics, 10 (1998), 1–46.
3. I. Daubechies, Ten Lectures on Wavelets, PA, SIAM and Philadelphia, 1992.
4. In-Keun Yu; Chang-Il Kim; Y. H. Song, A Novel Short-Term Load Forecasting Technique Using MODWT Analysis, Electric Machines and Power Systems, 24 (2000), 537–549.
5. J. Contreras, R.Espinola, F.J. Nogales, and A.J. Conejo, ARIMA models to predict next-day electricity prices, IEEE Transactions on Power Systems, 18(2005), 1014–20.
6. J.W. Hall, Adaptive selection of U.S, stocks with neural nets, in: G.J. Deboeck (Ed.), Trading on the Edge: Neural, Genetic and Fuzzy Systems for Chaotic Financial Markets, Willey, New York, 1994.
7. Philippe Masset, Analysis of Financial Time-Series Using Fourier and Wavelet Methods, University of Fribourg (Switzerland) - Faculty of Economics and Social Science, 2008.

8. P.Manchanda, J.Kumar, A.H.Siddiqi, Mathematical methods for modeling price fluctuations of financial time series, Journal of Franklin institute, 344(2007), 613–363.

9. R. Gencay, F. Seluk, and B. Whitcher, An Introduction to Wavelets and Other Filtering Methods in Finance and Economics, Academic Press, New York, 2002.

10. Rumaih M. Alrumaih and mohammad A. Al- Fawzan. Time series forecasting using wavelet denoising an application to Saudi stocks index, J.king Saudi Univ,14 (2002), 221–234.

11. SAS/ETS® 9.22 User's Guide, SAS Institute Inc, 2010.

12. S.A.M. Yaser, A.F. Atiya, Introduction to financial forecasting, Applied Intelligence, 6 (1996) 205–213.

13. S.J. Yao, Y.H. Song, L.Z. Zhang, X.Y. Cheng, MODWTand Networks for Short-term Electrical Load Forecasting, Energy Conversion and Management, 41(18) (2000), 1975–1988.

14. S.K. Aggarwal, L.M. Saini and Ashwani Kumar, Price forecasting using MODWTand LSE based mixed model in Australian electricity market. International Journal of Energy Sector Management, 2 (2008), 521–546.

15. Spyros Makridakis, Steven C. Wheelwright and Rob J. Hyndman. Forecasting methods and applications, Third edition, John Wiley& Sons, Inc, New York, 1998.

16. S.Ruey, analysis of financial time series. Second edition, John Wiley & Sons, Inc, 2005.

17. Todd Wittman, Time-Series Clustering and Association Analysis of Financial Data, CS 8980 Project, 2002.

18. Faisal Ababneh, S. Al Wadi and Mohd Tahir Ismail. Haar and Daubechies Wavelet Methods in Modeling Banking Sector. International Mathematical Forum, Vol. 8, 2013, no. 12, 551–566.

CIATION

S. Al Wadi, Abdulkareem Hamarsheh, Hazem Alwadi Maximum overlapping discrete wavelet transform in forecasting banking sector Applied Mathematical Sciences, Vol. 7, 2013, no. 80, 3995-4002 http://dx.doi.org/10.12988/ams.2013.36305

Series Methods for Forecasting Cocoa Bean Prices

K. Assis, A. Amran, Y. Remali and H. Affendy

[1]School of Sustainable Agriculture
[2]School of Science and Technology
[3]School of Business and Economics
[4]School of International Tropical
Forestry,Universiti Malaysia Sabah

ABSTRACT

The purpose of this study was to compare the forecasting performances of different time series methods for forecasting cocoa bean prices. The monthly average data of Tawau cocoa bean prices graded SMC 1B for the period of January 1992-December 2006 was used. Tawau is one of the top cocoa producers in the world along with the Ivory Coast, Ghana and Indonesia. Four different types of univariate time series methods or models were compared, namely the exponential smoothing, autoregressive integrated moving average (ARIMA), generalized autoregressive conditional heteroskedasticity (GARCH) and the mixed ARIMA/GARCH models. Root mean squared error (RMSE), mean absolute percentage error (MAPE), mean absolute error (MAE) and Theil's inequality coefficient (U-STATISTICS) were used as the selection criteria to determine the best forecasting model. This study revealed that the time series data were influenced by a positive linear trend factor while a regression test result showed the non-existence of seasonal factors. Moreover, the Autocorrelation function (ACF) and the Augmented Dickey-Fuller (ADF) tests have shown that the time series data was not stationary but became stationary after the first order of the differentiating process was carried out. Based on the results of the ex-post forecasting (starting from January until December 2006), the mixed ARI-

MA/GARCH model outperformed the exponential smoothing, ARIMA and GARCH models.Services Related Articles in ASCI Similar Articles in this Journal Search in Google Scholar View Citation Report Citation

INTRODUCTION

Cocoa, scientifically known as Theobroma cacao L. is the third-largest agricultural commodity in Malaysia after oil palms and rubber. Malaysia now exports cocoa products to sixty-six countries (Ministry of Plantation Industries and Commodities, 2006). Tawau is one of the top cocoa producers in Malaysia and even in the world along with the Ivory Coast, Ghana and Indonesia (Shanti, 2006). Domestic cocoa bean prices are changing from time to time and very volatile (Yusoff and Salleh, 1987; Arshad and Zainalabidin, 1994). Instability of cocoa prices creates significant risks to producers, suppliers, consumers and other parties that are involved in the marketing and production of cocoa beans, particularly in Malaysia. In risky conditions and amidst price instability, forecasting is very important in helping to make decisions. Accurate price forecasts are particularly important to facilitate efficient decision making as there is time lag intervenes between making decisions and the actual output of the commodity in the market.

Modelling or forecasting of agricultural price series, like that of other economic time series, has traditionally been carried out either by building an econometric model or by applying techniques developed for analyzing stationary time series. Time series forecasting is a major challenge in many real world applications such as stock price analysis, palm oil prices, natural rubber prices, electricity prices and flood forecasting. This type of forecasting is to predict the values of a continuous variable (called as response variable or output variable) with a forecasting model based on historical data. There are two types of time series forecasting modeling methods; univariate and multivariate. Univariate modeling methods generally used time only as an input variable with no other outside explanatory variables (Celia et al., 2003). This forecasting method is often called univariate time series modeling. A few commonly employed methods in univariate time series models are exponential smoothing, autoregressive-integrated-moving aver-age (ARIMA) and Autoregressive Conditional Heteroscedastic (ARCH) (Kahforoushan et al., 2010).

The last few decades have witnessed significant advances in the topic of exponential smoothing. It has established itself as one of the lead-ing forecasting strategies (Robert and Amir, 2009). Fatimah and Ro-slan (1986) confirmed the suitability of univariate ARIMA models in agricultural prices forecasting. Shamsudin et al. (1992) has noted that ARIMA models have the advantage of relatively low research costs when compared with econometric models, as well as efficiency in short term forecasting. One of the earliest time series models allow-ing for heteroscedasticity is the Autoregressive Conditional Heterosce-dastic (ARCH) model introduced by Engel (1982). Bollerslev (1986) extended this idea into Generalized Autoregressive Conditional Het-eroscedastic (GARCH) models which give more parsimonious results than ARCH models, similar to the situation where ARMA models are preferred over AR models. Kamil and Noor (2006) have developed a time series model of Malaysian palm oil prices by using ARCH mod-els. Zhou et al. (2006) have proposed a new network traffic prediction model based on non-linear time series ARIMA/GARCH. They found that the proposed ARIMA/GARCH outperformed the existing Fraction-al Autoregressive Integrated Moving Average (FARIMA) model in terms of prediction accuracy. Therefore, the objective of this research was to compare the forecasting performances of four different univariate time series methods or models for forecasting cocoa bean prices (i.e., Tawau cocoa bean prices), namely exponential smoothing, ARIMA, GARCH and the mixed ARIMA/GARCH models.

MATERIALS AND METHODS

The monthly Tawau cocoa bean prices graded SMC 1B was used for this study which was collected from the official website of The Malaysian Cocoa Board (http://www.koko.gov.my/lkmbm/loader.cfm?page=statisticsFrm.cfm). The time series data was measured in Ringgit Malaysia per tonne (RM/tonne). The time series data ranged from January 1992 until December 2006. The coefficient of variation (V) was used to measure the index of instability of the time series data. The coefficient of variation (V) is defined as:

$$V = \frac{\sigma}{\overline{Y}}$$

where σ is the standard deviation and

$$\overline{Y} = \frac{\sum\limits_{t=1}^{n} Y_t}{n}$$

is the mean of Tawau cocoa bean prices changes.

A completely stable data has $V = 1$, but unstable data are characterized by a $V > 1$ (Telesca et al., 2008).

Regression analysis : was used to test whether trends and seasonal factors exist in the time series data. The existence of linear trend factors was tested through this regression equation

$$Y = \beta_0 + \beta_1 \text{Trend} + \varepsilon \qquad \varepsilon \sim WN\left(0, \sigma^2\right)$$

with Y is the time series data of the study, Trend is the linear trend factor, $\beta 0$ and $\beta 1$ are parameters and ε is the error of the model with an assumption of White Noise (WN). The hypothesis of the model was

- H0: $\beta 1$ = (Non-existence of linear trend factor)
- H1: $\beta 1 \neq$ (Linear trend factor exists)

With the month of January as the base month, the existence of seasonal factor was detected by using regression as shown below:

$$Y = \beta_0 + \beta_1 Trend + \beta_2 Feb + \beta_3 Mar + \beta_4 Apr + \beta_5 May + \beta_6 Jun + \beta_7 Jul$$
$$+\beta_8 Aug + \beta_9 Sep + \beta_{10} oct + \beta_{11} Nov + \beta_{12} Dec + \varepsilon$$

and hypothesis was defined as:

- H0: $\beta2 = \beta3 = \beta4 = ...\beta12 = 0$ (Non-existence of seasonal factor)
- H1: At least one of $\beta2, \beta3, ..., \beta12 \neq 0$ (Seasonal factor exists)

The correlogram and Augmented Dickey-Fuller (ADF) test were chosen to test the stationary of the time series data.

Exponential Smoothing

The h-periods-ahead forecast is given by:

$$\hat{Y}_{t+h} = a + bh$$

with a and b are permanent components. Both of these parameters are counted by the following equations

$$a_t = \alpha Y_t + (1-\alpha)(a_{t-1} + b_{t-1})$$

$$b_t = \beta(a_t - a_{t-1}) + (1-\beta)b_{t-1}$$

with $0<\alpha, \beta<1$.

ARIMA

This study followed the Box-Jenkins methodology which involves four steps. These are identification, estimation, model checking and forecasting. ARMA (p, q) processes can be simply expressed as the following two Eq.

$$Y_t = X_t \gamma + e_t \tag{1}$$

$$e_t = \sum_{i=1}^{p} \phi_i u_{t-1} + \sum_{j=1}^{q} \theta_j \varepsilon_{t-j} + \varepsilon_t \tag{2}$$

where, xt is the explanatory variables, etis the disturbance term, εt is the innovation in the disturbance, p is the order of AR term, q: the order of MA term. In Eq. 2, the disturbance term (μt) again consists of three parts. The first part is AR terms and the second part is MA terms. The last one is just a white-noise innovation term. If we replace the data (Y) with the difference data ($\Delta yt = Yt-Yt-1$), then the ARMA models become ARIMA(p, d, q) models.

GARCH

The standard form of GARCH(p,q) models can be specified as following three equations:

$$Y_t = X_t \gamma + \varepsilon_t \tag{3}$$

$$\varepsilon_t = v_t \sqrt{\sigma_t^2} \tag{4}$$

$$\sigma_t^2 = \delta + \sum_{i=1}^{q} \alpha_i \varepsilon_{t-i}^2 + \sum_{j-1}^{p} \beta_j \sigma_{t-j}^2 \tag{5}$$

where, p is the order of GARCH term, q is the order of ARCH term and σ2v. Equation 3 and 5 are called mean equation and conditional variance equation, respectively. The mean equation is written as a function of exogenous variables (xt) with an error term (μt). The variance equation is a function of mean (δ), ARCH (μ2t-i) and GARCH term (μ2t-t).

ARIMA/GARCH

Combination of ARIMA(p,d,q) and GARCH(p,q) are written as below:

$$\left(\Delta Y_t\right)^d = \sum_{i=1}^{p}\phi_i\left(\Delta Y_{t-i}\right)^d + \varepsilon + \sum_{i=1}^{q}\theta_j\varepsilon_{t-j} \qquad \varepsilon_t \sim WN\left(0,\sigma_t^2\right)$$

(6)

$$\sigma_t^2 = \delta + \sum_{j=1}^{q}\beta_j\varepsilon_{t-j}^2 + \sum_{i=1}^{p}\alpha_i\sigma_{t-i}^2$$

(7)

Eight model selection criteria as suggested by Ramanathan (2002) were used to chose the best forecasting models among ARIMA and GARCH models (Table 1). While, the best time series methods for forecasting Tawau cocoa bean prices was chosen based on the values of four criteria, namely RMSE, MAE, MAPE and U-statistics (Table 2). Finally, the selected model was used to perform short-term forecasting for the next twelve months for Tawau cocoa bean prices starting from January 2007 until December 2007.

RESULTS

The results showed that the coefficient of variation (V) of the time series data was 1.012 (V>1). Because of the V value was closed to 1, so this study was concluded that the time series data was stable (Telesca et al., 2008). The results of the regression analysis have shown that positive linear trend factor exists in the time series data but seasonal factor was not. With referring to the correlogram and the Augmented Dickey-Fuller tests results, the time series data of the study was not stationary. But after the first order of differencing was carried out, the time series data became stationary (Fig. 1).

Exponential Smoothing

The double exponential smoothing method was used as the regression result has showed the positive linear trend factor exists in the time

Table 1: Model Selection Criteria (Ramanathan, 2002)

Criteria	Formula
Alc FPE GCV HQ RICE SCHWARZ SGMASQ SHIBATA	$\left(\dfrac{ESS}{n}\right)e^{2f/n}$ $\left(\dfrac{ESS}{n}\right)\dfrac{n+f}{n-f}$ $\left(\dfrac{ESS}{n}\right)\left[1-\left(\dfrac{f}{n}\right)\right]^{-2}$ $\left(\dfrac{ESS}{n}\right)(1nn)^{2f/n}$ $\left(\dfrac{ESS}{n}\right)\left[1-\left(\dfrac{2f}{n}\right)\right]^{-1}$ $\left(\dfrac{ESS}{n}\right)n^{f/n}$ $\left(\dfrac{ESS}{n}\right)\left[1-\left(\dfrac{f}{n}\right)\right]^{-1}$ $\left(\dfrac{ESS}{n}\right)\dfrac{n+2f}{n}$

Yt: The actual value at time t, : The forecast value at time t, n: The number of observations; ESS: The error sum of square

series data. Double exponential smoothing models consist with two parameters which symbolized as α for mean and β for trend. The best model of the double exponential smoothing has been selected based on the lowest value of MSE (Mean Square Error) from combination of α and β with condition $0<\alpha, \beta<1$.

Table.2: Forecast accuracy criteria

Criteria	Formula
RMSE	$\sqrt{\dfrac{ESS}{n}}$
MAE	$\sum\limits_{t=1}^{n}\left|Y_t - \hat{Y}_t\right|$
MAPE	$\dfrac{\sum\limits_{t=1}^{n}\left|\dfrac{Y_t - \hat{Y}_1}{Y_t}\right|}{}\times 100\%$
U-statistics	$\sqrt{\sum\limits_{t=1}^{n}\hat{Y}^2 / n} + \sqrt{\sum\limits_{t=1}^{n}Y^2 / n}$

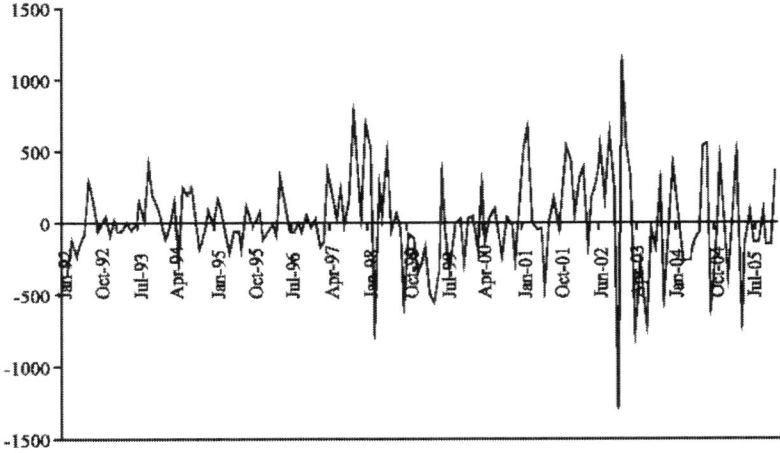

Figure.1: Time series data (after first order of differencing).

Table 3: Error Sum of Square (ESS) according to α and β values

a / p	0.1	0.2	04	0.4	0.5	06	07	08	0.9
01	173185627	70362670	41458913	30011334	24384697	21244412	19366607	18231486	17589723
82	171444565	61358397	36273998	27053389	22754641	20391822	19006661	18223235	17866015
OS	138700812	53824940	32336246	23412232	22160211	20307532	19216E63	18673826	183203_4
OA	130053979	46³89161	30069608	24960672	22293468	20684210	19769690	19377921	19415785
OS	130417274	41201810	23317287	25242334	22771650	21264717	20458721	20212951	20143538

Table 4: EViews output of the double exponential smoothing model

Sample: 1992M01 2005M12	
included observations: 168	
Method: Holt-Winters No Seasonal	
Original Series: TAWAU	
Forecast series TAWAUSM	

Analysis	Values
Parameter	0.9000
Alpha	0.1000
Beta	17589723
Sum of squared residual	323.5749
Root mean squared error	
End of period levels	
Mean	4852.068
Trend	−23.43109

The result showed that combination $\alpha = 0.9$ and $\beta = 0.1$ was the best forecasting model of double exponential smoothing method (Table 3). The double exponential smoothing model was written in equation form as (Table 4).

$$F_{T+h} = a + bh = 4852.068 + (h)*(-23.43109)$$

ARIMA

All models which fulfilled the criteria of p+q≤5 have been considered and compared in this study and there were twenty ARIMA (p, d, q) models which fulfilled the criteria. Parameters of the models were estimated with the least square method. Parameters which were not significant at 5% confidence level were dropped from the model. Using the eight model selection criteria suggested by Ramanathan (2002), the ARIMA (3, 1, 2) model was selected as the best model among the other ARIMA models. However, the parameters of AR (1) and MA (1) were found not significant and thus dropped from the model.

Table 5: Estimation of ARIMA (3, 1, 2)

Variables	Coefficient	Standard error	Z-statistic	p-value
Constant	18.130660	26.185420	0.692395	0.4897
	−0.895200	0.051784	−7.287200	<0.0001"
	0.118147	0.039606	2.983055	0.0033*
MA(2)	0.936905	0.045159	20.746770	<0.0001"

*$p<0.05$

Table 6: Estimation of GARCH (1, 1)

	Mean equation			
Variables	Coefficient	Standard error	Z-statistic	p-value
Constant	2.2945	17.5522	0.1307	0.8960
Conditional variance equation				
Constant	5896.95	2841.67	2.0752	0.0380*
en	0.2753	0.1009	23286	0.0064*
ci^2	0.6887	0.0961	7.1633	<0.0001*

*$p<0.05$

Table 7: Estimation of ARIMA (3, 1, 2)/GARCH (1, 1)

Variables	Coefficient	Standard error	Z-statistic	p-vahie
ARINIA (3, 1, 2)				
Constant	10.42818	19.2344	0.542163	0.5877
AR (2)	−0.86716	0.050758	-17.08420D	<0.0001*
AR (3)	0.087526	0.036952	2.368676	0.0179*
MA (2)	0.944859	0.030012	31.483150	<0.0001*
GARCH (1, 1)				
C	4606.362	2778.705	1.657737	0.0974
et.L	0.237459	0.099897	2.377032	0.0175*
02.,	0.736343	0.106755	6.897485	<0.0001*

*p<0.05
The ARIMA (3, 1, 2) model was written in equation form as (Table 5).

GARCH

Identification and estimation of GARCH (p, q) models in this study were done by following the four steps that were ARCH effect checking, estimation, model checking and forecasting. Four GARCH (p,q) models were selected and compared, namely GARCH (1, 1), GARCH (1, 2), GARCH (2, 1) and GARCH (2, 2). Using the eight model selection criteria suggested by Ramanathan (2002), the GARCH (1, 1) model has been selected as the best model among the other three GARCH models.

The GARCH(1,1) model was written in equation form as (Table 6):

$$\hat{Z}_t = 2.2945 \qquad \text{(Mean equation)}$$

$$\hat{\sigma}_t^2 = 5896.95 + 0.2753\varepsilon_{t-1}^2 + 0.6887\sigma_{t-1}^2 \qquad \text{(Conditional variance equation)}$$

ARIMA/GARCH

ARCH effect which was tested by using a regression analysis exists in the ARIMA (3, 1, 2) model. That means the ARIMA (3, 1, 2) model could be mixed with the best GARCH model (i.e., GARCH(1, 1)).

The ARIMA(3, 1, 2)/GARCH(1, 1) model was written in equation form as (Table 7):

$$\hat{Z}_t = 10.42818 - 0.86716z_{t-2} + 0.087526z_{t-3} + 0.944859\varepsilon_{t-1}$$

$$\hat{\sigma}_t^2 = 4606.362 + 0.237459\varepsilon_{t-1}^2 + 0.736343\sigma_{t-1}^2$$

Table 8: Four model selection criteria

Criteria	Double exponatal SMOOthillft	AR1MA (3, 1.2)	GARCH (1. 1)	ARJIMA (3. 1.2y GARCH (1, I)
RMSE	324.43006	183.3087257	15118800832	153.3006431
MAE	292.65075	144.9835	12280825	126.74
MAPS	5.805836015	2.936916969	2.417800798	2.552741574
U-8tatistics	0.033466951	0.018211057	0.01604509	0.015509933

Model Selection

Four model selection criteria were used to select the best forecasting model from the four different types of time series methods. Based on the results of the ex-post forecasting (starting from January until December 2006), the ARIMA (3, 1, 2)/GARCH (1, 1) model was the best short-term forecasting model of Tawau cocoa bean price graded SMC 1B (Table 8).

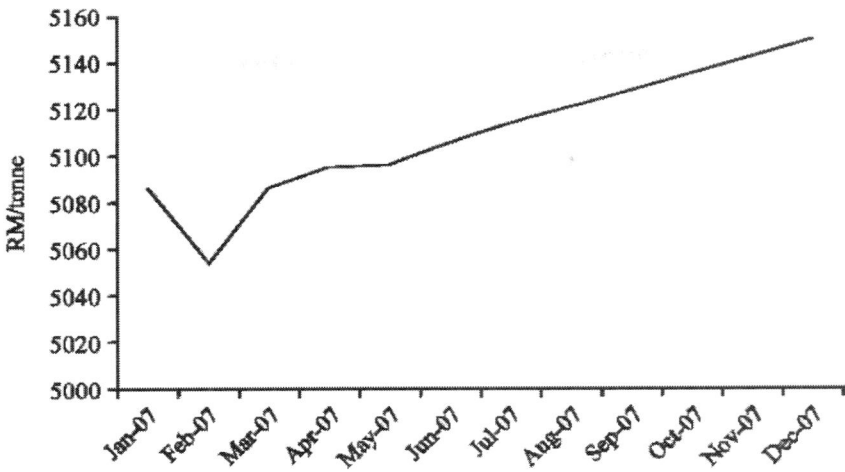

Figure.2: Short-term forecasting of tawau cocoa bean prices.

Ex-Ante Forecasting

Based on the ex-ante forecasting by using the mixed ARIMA/GARCH model, Fig. 2 shows that the short-term forecasting indicated an upward trend of Tawau cocoa bean prices for the period January-December 2007.

DISCUSSION

The result showed that the time series data (starting January 1992 until December 2006) was stable. This is contradict with the previous researches (Yusoff and Salleh, 1987; Arshad and Zainalabidin, 1994) which stated that domestic cocoa bean prices are changing from time to time and very volatile. The results of the regression analysis have shown that positive linear trend factor exists in the time series data but seasonal factor was not. That means the cocoa bean prices of Tawau have increased in the period of 1992-2006 but seasonal factor which is usually related to climate change has not given any significant influence on the monthly changes of cocoa bean prices. The mixed ARIMA/

GARCH model outperformed the exponential smoothing, ARIMA and GARCH for the case of forecasting monthly Tawau cocoa bean prices. This is in agreement with the findings in the literature (Zhou et al., 2006). Some of previous research have found that ARIMA models (Fatimah and Roslan, 1986; Shamsudin et al., 1992; Kahforoushan et al., 2010) and also GARCH-type models (Kamil and Noor, 2006) were the best or suitable price forecasting models in terms of prediction accuracy, but the accuracy of the mixed ARIMA/GARCH should also be considered in price forecasting for the future researches.

CONCLUSION

This study investigates four different types of univariate time series methods, namely exponential smoothing, ARIMA, GARCH and the mixed ARIMA/GARCH. The results showed that the mixed ARIMA/GARCH model outperformed the exponential smoothing, ARIMA and GARCH for forecasting Tawau cocoa bean prices. Forecasting the future prices of cocoa bean through the most accurate univariate time series model can help the Malaysian government as well as the buyers (e.g., exporters and millers) and sellers (e.g., farmers and dealers) in cocoa bean industry to perform better strategic planning and also to help them in maximizing revenue and minimizing the cost of price.

REFERENCES

1. Arshad, F.A. and M. Zainalabidin, 1994. Price discovery through crude palm oil futures market: An economic evaluation. Proceedings of the 3rd Annual Congress on Capitalising the Potentials of Globalisation-Strategies and Dynamics of Business, (CPGSDB'94), IMDA, Malaysia, pp: 73–92.
2. Bollerslev, T., 1986. Generalized autoregressive conditional heteroskedasticity. J. Econ., 31: 307–327.
3. Celia, F., G. Ashish, R. Amar and S. Les, 2003. Forecasting women's apparel sales using mathematical modeling. Int. J. Clothing Sci. Technol., 15: 107–125.
4. Engle, R.F., 1982. Autoregressive conditional heteroscedasticity with estimates of the variance of United Kingdom inflation. Econometrica, 50: 987–1007.
5. Fatimah, M.A. and A.G. Roslan, 1986. Univariate approach towards cocoa price forecasting. Malaysian Agric. Econ., 3: 1–11.
6. Kahforoushan, E., M. Zarif and E.B. Mashahir, 2010. Prediction of added value

of agricultural subsections using artificial neural networks: Box-Jenkins and holt-winters methods. J. Dev. Agric. Econ., 2: 115–121.

7. Kamil, A.A. and A. Md. Noor, 2006. Time series modeling of Malaysian raw palm oil price: Autoregressive Conditional Heteroskedasticity (ARCH) model approach. Discov. Math., 28: 19–32.
8. Ministry of Plantation Industries and Commodities, 2006. Report on Malaysia plantation industries and commodities (2001-2005). Putrajaya.
9. Ramanathan, R., 2002. Introductory Econometric with Applications. 5th Edn., Thomson Learning, South Carolina, pp: 688.
10. Robert, R.A. and F.A. Amir, 2009. A new bayesian formulation for holt's exponential smoothing. J. Forecast., 28: 218–234.
11. Shamsudin, M.N., M.L. Rosdi and T.C. Ann, 1992. An econometric analysis of cocoa prices: A structural approach. J. Ekonomi Malaysia, 25: 3–17.
12. Shanti, G., 2006. Destination cocoa town. COPAL Cocoa Info, A Weekly Newsletter of Cocoa Producers Alliance, Cocoa Producers Alliance, Issue No. 180. May 22-26. http://www.copal-cpa.org/newsletters/Newsletter%20No%20180.pdf.
13. Telesca, L., M. Bernardi and C. Rovelli, 2008. Time-scaling analysis of lightning in Italy. Commun. Nonlinear Sci. Numer. Simul., 13: 1384–1396.
14. Yusoff, M. and M. Salleh, 1987. The elasticities of supply and demand for malaysian primary commodity exports. Malaysian J. Agric. Econ., 4: 59–72.
15. Zhou, B., D. He and Z. Sun, 2006. Modeling and Simulation Tools for Emerging Telecommunication Networks. Springer, USA.

CITATION

K. Assis, A. Amran, Y. Remali and H. Affendy A Comparison of Univariate Time Series Methods for Forecasting Cocoa Bean Prices DOI: 10.3923/tae.2010.207.215

Malaysian Gold Prices Modelling and Forecasting

Nor Hamizah Miswan, Pung Yean Ping, and Maizah Hura Ahmad

Department of Mathematical Sciences, Faculty of Science Universiti Teknologi Malaysia, 81310 UTM Skudai, Johor, Malaysia

ABSTRACT

In developing a time series model, parameter estimation is one of the crucial steps. Common methods of estimation include method of moment (MME), ordinary least square estimation (OLS) and maximum likelihood estimation (MLE). The purpose of the current study is to model and forecast the prices of Malaysian gold called kijang emas using Box-Jenkins methodology. To find the best model, parameter estimates using OLS and MLE were computed. Based on the Akaike information criteria (AIC) and mean absolute percentage error (MAPE), the model estimated with OLS was found to perform better.

INTRODUCTION

Kijang Emas is the official Malaysian gold bullion coin minted by the Royal Mint of Malaysia. Kijang emas prices are volatile with huge price swings. Volatility refers to a condition where the conditional variance changes between extremely high and low values. The forecasting of kijang emas prices is useful for investment purposes in Malaysia.

In developing any time series model, parameter estimation is one of the crucial steps. Common methods of estimation include method of moment (MME), ordinary least square estimation (OLS) and maximum likelihood estimation (MLE). MLE for example, was used to estimate ARFIMA models [1]. MME on the other hand is rarely used in time series analyses because it produces poor estimates [2]. Although it is easy, MME is not an efficient estimation method for ARIMA model because it works for only Autoregressive models of large sizes.

The purpose of the current study is to model and forecast Kijang emas prices using Box Jenkins methodology. However, the focus of this paper is parameter estimation.

METHODOLOGY

The Box-Jenkins methodology refers to a several-step process for identifying, fitting, and checking ARIMA models with time series data. Forecasts are then made using the fitted model. In describing time series, Autoregressive Moving Average (ARMA (p,q)) model is used when the series are stationary while the Autoregressive Integrated Moving Average (ARIMA (p,d,q)) process is used when the series are nonstationary. While a nonstationary in mean time series can be made stationary by differencing (d), variance stabilizing transformation known as Box-Cox can be used to reduce anomalies such as non-additivity, non-normality and heteroscedasticity [3].

In time series analysis, the most essential steps are to identify and build a model of the available data. To describe the model identification, consider the general ARIMA(p,d,q) process,

$$(1 - \emptyset_1 B - \cdots - \emptyset_p B^p)(1 - B)^d y_t = \delta + (1 - \theta_1 B - \cdots - \theta_q B^q) a_t$$

Model identification refers to the identification of the required transformation and the orders of p and q for the model. The next step is to estimate the parameters of the selected models. The current study investigates two methods of parameter estimation, namely MLE and OLS.

Maximum Likelihood Estimation (MLE)

The idea of maximum likelihood estimator is to determine the parameters that maximize the probability or likelihood of the sample data.

The general procedures are as follows:

(i) Let Y be a continuous random variable with probability density function

$$f(y; \theta_1, \theta_2, \ldots, \theta_k)$$

where $\theta_1, \theta_2, \ldots, \theta_k$ are k unknown constant parameters that need to be estimated. The likelihood function is given by the following product:

$$L(y_1, y_2, \ldots, y_n | \theta_1, \theta_2, \ldots, \theta_k) = \prod_{i=1}^{n} f(y_i; \theta_1, \theta_2, \ldots, \theta_k)$$

$$L(y_i | \theta_j) = \prod_{i=1}^{n} f(y_i; \theta_j)$$

where i=1,2,...,n and j=1,2,.../k

(ii) The logarithmic likelihood function is obtained by taking natural logarithms:

$$l = \ln L = \ln \prod_{i=1}^{n} f(x_i; \theta_j) = \sum_{i=1}^{n} \ln f(x_i; \theta_j)$$

(iii) To solve for θ_j, differentiate with respect to θ_j and set the derivative to zero

$$\frac{dl}{d\theta_j} = \frac{d}{d\theta_j} \sum_{i=1}^{n} \ln f(x_i; \theta_j) = 0$$

The parameter estimates can also be computed by maximizing the log-likelihood function.

To estimate a Box-Jenkins model, consider the general ARMA (p,q) model,

$$y_t = \emptyset_1 y_{t-1} + \cdots + \emptyset_p y_{t-p} + \delta + a_t - \theta_1 a_{t-1} - \cdots - \theta_q a_{t-q}$$

$$(1)$$

Suppose that $a_t \sim N(0, \sigma_a^2)$ is white noise. Rewriting equation (1) results in,

$$a_t = \theta_1 a_{t-1} + \cdots + \theta_q a_{t-q} + y_t - \delta - \emptyset_1 y_{t-1} - \cdots - \emptyset_p y_{t-p}$$

$$(2)$$

The probability distribution function of a_t is given by,

$$f(a_t) = \frac{1}{\sqrt{2\pi\sigma_a^2}} \exp\left\{\frac{-a_t^2}{2\sigma_a^2}\right\}$$

$$(3)$$

The likelihood function is given by the joint probability density function of a_t,

$$L = \prod_{t=1}^{n}\left(\frac{1}{\sqrt{2\pi\sigma_a^2}} \exp\left\{\frac{-a_t^2}{2\sigma_a^2}\right\}\right)$$

Take natural logarithm to form log-likelihood function,

$$l = \ln L = \ln \prod_{t=1}^{n} \left(\frac{1}{\sqrt{2\pi\sigma_a^2}} \exp\left\{ \frac{-a_t^2}{2\sigma_a^2} \right\} \right)$$

$$= \sum_{t=1}^{n} \ln \left(\frac{1}{\sqrt{2\pi\sigma_a^2}} \exp\left\{ \frac{-a_t^2}{2\sigma_a^2} \right\} \right)$$

$$= \sum_{t=1}^{n} \left(\ln\left\{ \frac{1}{2\pi\sigma_a^2} \right\}^{\frac{1}{2}} + \left\{ \frac{-a_t^2}{2\sigma_a^2} \right\} \right)$$

$$= \sum_{t=1}^{n} \left(-\frac{1}{2} \ln\{2\pi\sigma_a^2\} - \frac{a_t^2}{2\sigma_a^2} \right)$$

$$= -\frac{n}{2} \ln\{2\pi\sigma_a^2\} - \frac{\sum_{t=1}^{n} a_t^2}{2\sigma_a^2}$$

(4)

Where

$$a_t = \theta_1 a_{t-1} + \cdots + \theta_q a_{t-q} + y_t + \varnothing_1 y_{t-1} - \cdots - \varnothing_p y_{t-p}$$

Based on the assumption that y_t is stationary and $\{a_t\} \sim$ iid $N(0,\sigma_a^2)$ random variables, y_t is replaced by the sample mean $\bar{y}$ and a_t by its expected value which is zero. From (2), it can be assumed that $a_t = a_{t-1} = \cdots a_{t-q} = 0$ and a_t is calculated by using (3) for $t \geq (p+1)$.

After obtaining the parameter estimates θ_i and ϕ_i, the estimate σ_a^2 is calculated from,

$$\hat{\sigma}_a^2 = \frac{\sum_{t=p+1}^{n} a_t^2}{df}$$

where $df = (n-p) - (p+q+1) = n - (2p+q+1)$ which is the number of terms used in $\sum_{t=p+1}^{n} a_t^2$ minus the number of parameters estimated.

Ordinary Least Squares Estimation (OLS)

The idea of Least Squares Estimation is to find the parameters that minimize the sum of squared errors. The general procedures are as follows:

(i) Let y_t be the actual value of the data $\hat{y}_t$ and be the fitted value. Take the sum of squared errors as follows,

$$S = \sum_{t=1}^{n} e_t^2 = \sum_{t=1}^{n} (y_t - \hat{y}_t)^2$$

Supposed that $\hat{y}_t$ contain the parameters to be estimated. Let β_i be the estimators. Hence, the sum of squared errors can be written as

$$S = \sum_{t=1}^{n} (y_t - \hat{y}_t(\beta_i))^2$$

(5)

(ii) Equation (5) can be minimized by taking the derivative with respect to β_i and set the derivative to zero. Then solve for β_i.

$$\frac{dS}{d\beta_i} = \frac{d}{d\beta_i} \sum_{t=1}^{n} (y_t - \hat{y}_t(\beta_i))^2 = 0$$

To estimate a Box-Jenkins model, consider the general ARMA (p,q) model,

$$y_t = \emptyset_1 y_{t-1} + \cdots + \emptyset_p y_{t-p} + \delta + a_t - \theta_1 a_{t-1} - \cdots - \theta_q a_{t-q} \tag{6}$$

Suppose that $a_t \sim N(0, \sigma_a^2)$ is white noise. Rewriting equation (6) results in,

$$a_t = \theta_1 a_{t-1} + \cdots + \theta_q a_{t-q} + y_t - \delta - \emptyset_1 y_{t-1} - \cdots - \emptyset_p y_{t-p}$$

$$a_t = \sum_{j=1}^{q} \theta_j a_{t-j} + y_t - \delta - \sum_{i=1}^{p} \emptyset_i y_{t-i} \tag{7}$$

Hence, the sum of squared errors is given by the following equation,

$$S = \sum_{t=1}^{n} a_t^2 = \sum_{t=1}^{n} \left(\sum_{j=1}^{q} \theta_j a_{t-j} + y_t - \delta - \sum_{i=1}^{p} \emptyset_i y_{t-i} \right)^2 \tag{8}$$

Differentiate (8) with respect to δ, ϕ_i and θ_j and set the derivatives to zero,

$$\frac{dS}{d\delta} = 0$$

$$\frac{dS}{d\delta} = -2 \sum_{t=1}^{n} \left(\sum_{j=1}^{q} \theta_j a_{t-j} + y_t - \delta - \sum_{i=1}^{p} \emptyset_i y_{t-i} \right)^2 = 0$$

$$\sum_{t=1}^{n} \left(\sum_{j=1}^{q} \theta_j a_{t-j} + y_t - \delta - \sum_{i=1}^{p} \emptyset_i y_{t-i} \right) = 0$$

$$\sum_{t=1}^{n} \sum_{j=1}^{q} \theta_j a_{t-j} + \sum_{t=1}^{n} y_t - n\delta - \sum_{t=1}^{n} \sum_{i=1}^{p} \emptyset_i y_{t-i} = 0$$

$$n\delta + \sum_{t=1}^{n} \sum_{i=1}^{p} \emptyset_i y_{t-i} - \sum_{t=1}^{n} \sum_{j=1}^{q} \theta_j a_{t-j} = \sum_{t=1}^{n} y_t \tag{9}$$

$$\frac{dS}{d\emptyset_i} = 0$$

$$\frac{dS}{d\emptyset_i} = -2\sum_{t=1}^{n}\sum_{i=1}^{p} y_{t-i}\left(\sum_{j=1}^{q}\theta_j a_{t-j} + y_t - \delta - \sum_{i=1}^{p}\emptyset_i y_{t-i}\right) = 0$$

$$\sum_{t=1}^{n}\sum_{i=1}^{p} y_{t-i}\left(\sum_{j=1}^{q}\theta_j a_{t-j} + y_t - \delta - \sum_{i=1}^{p}\emptyset_i y_{t-i}\right) = 0$$

$$n\delta\sum_{i=1}^{p} y_{t-i} + \emptyset_i\sum_{t=1}^{n}\left(\sum_{i=1}^{p} y_{t-i}\right)^2 - \theta_j\sum_{t=1}^{n}\left(\sum_{i=1}^{p} y_{t-i}\right)\left(\sum_{j=1}^{q} a_{t-j}\right) = \sum_{t=1}^{n}\sum_{i=1}^{p} y_t y_{t-i}$$

(10)

$$\frac{dS}{d\theta_j} = 0$$

$$\frac{dS}{d\theta_j} = 2\sum_{t=1}^{n}\sum_{j=1}^{q} a_{t-j}\left(\sum_{j=1}^{q}\theta_j a_{t-j} + y_t - \delta - \sum_{i=1}^{p}\emptyset_i y_{t-i}\right) = 0$$

$$\sum_{t=1}^{n}\sum_{j=1}^{q} a_{t-j}\left(\sum_{j=1}^{q}\theta_j a_{t-j} + y_t - \delta - \sum_{i=1}^{p}\emptyset_i y_{t-i}\right) = 0$$

$$\theta_j\sum_{t=1}^{n}\left(\sum_{j=1}^{q} a_{t-j}\right)^2 + \sum_{t=1}^{n}\sum_{j=1}^{q} y_t a_{t-j} - n\delta\sum_{j=1}^{q} a_{t-j}$$

$$- \emptyset_i\sum_{t=1}^{n}\left(\sum_{i=1}^{p} y_{t-i}\right)\left(\sum_{j=1}^{q} a_{t-j}\right) = 0$$

$$n\delta\sum_{j=1}^{q} a_{t-j} + \emptyset_i\sum_{t=1}^{n}\left(\sum_{i=1}^{p} y_{t-i}\right)\left(\sum_{j=1}^{q} a_{t-j}\right) - \theta_j\sum_{t=1}^{n}\left(\sum_{j=1}^{q} a_{t-j}\right)^2 = \sum_{t=1}^{n}\sum_{j=1}^{q} y_t a_{t-j}$$

(11)

Equations (9), (10) and (11) are solved simultaneously to obtain δ, ϕ_i and θ_j.

Performances Evaluation

The modeling and forecasting performances of models will be evaluated by using Akaike's information criterion (AIC) and mean absolute percentage error (MAPE) respectively as follows:

$$AIC = n \ln MSE - n \ln n + 2p$$

$$MAPE = \left\{ \left[\sum_{t=1}^{n} \left| \frac{y_t - \hat{y}_t}{y_t} \right| \right] / n \right\} \times 100\%$$

DATA ANALYSIS

The data used in this study are the Kijang emas prices recorded from 18th July 2001 until 25th September 2012 as plotted in Figure 1.

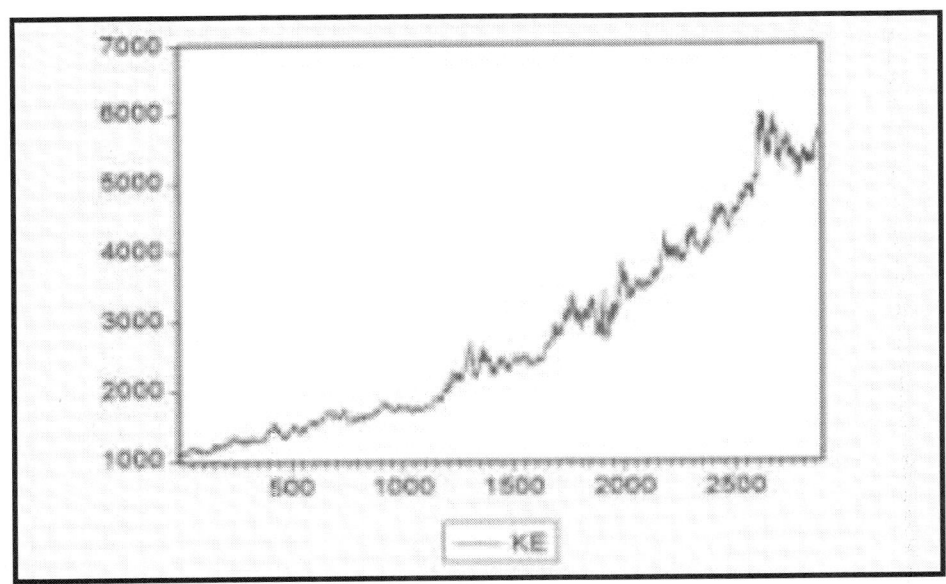

Figure 1: Kijang Emas Prices.

The analyses start with stationarity testing, followed by the process of model identification and parameter estimation. The nonstationarity of the series are first removed by using Box-Cox transformation, followed by differencing.

Consider the following equation of ARIMA(p,d,q) model,

$$\emptyset_p(B)(1-B)^d y_t = \delta + \theta_q(B)a_t$$

(12)

where

$\phi_p(B) = 1 - \phi_1 B - \cdots - \phi_p B^p$, the autoregressive operator of order p

$\theta_q(B) = 1 - \theta_1 B - \cdots - \theta_q B^q$ the moving average operator of order q $(1-B)^d$ is the dth difference, B is backward shift operator and a_t is the error term at time t.

The values of parameters that are to be estimated from equation (12) are δ, ϕ_p and θ_q depending on the values of p and q.

In order to obtain the values of parameters by using MLE, R software was used.

Five ARIMA(p,d,q) models were developed: ARIMA(1,1,1), ARIMA(2,1,1), ARIMA(1,1,0), ARIMA(2,1,0) and ARIMA(0,1,1). The corresponding AIC values are 23975.87, 23978.44, 23977.11, 23976.45 and 23976.52 respectively. In order to obtain the values of parameters by using OLS, a software called Eviews was used. Five ARIMA (p,d,q) models were also developed: ARIMA(1,1,1), ARIMA(2,1,1), ARIMA(1,1,0), ARIMA(2,1,0) and ARIMA(0,1,1). Their corresponding AIC values are 10.08545, 10.08741, 10.08901, 10.08816 and 10.08867 respectively. The AIC values for ARIMA(1,1,1) were the lowest for models estimated by both MLE and OLS.

Forecasting

Based on the values of AIC for MLE and OLS methods, ARIMA(1,1,1) is the best model. The models are used for forecasting and the performances are evaluated by computing MAPE. The results are tabulated in Table 1 as follows:

Table 1: Forecasting performances of ARIMA (1,1,1)

Model	Parameter Estimation Method	MAPE
ARIMA(1,1,1)	Maximum Likelihood Estimation	2.860522
	Ordinary Least Squares Estimation	0.812356

From Table 1, the MAPE value for OLS is smaller than the MAPE value for MLE. Hence, it can be concluded that OLS is the best estimation method of ARIMA model for kijang emas prices data.

CONCLUSIONS

For kijang emas prices data, the method of OLS gives better forecasts as the MAPE value is smaller than the MAPE value for MLE method. This is because the graph of kijang emas prices as plotted in Figure 1 tends to be linear. From the figure, it is clear that there is an increasing trend throughout the time. As described by Hutcheson [4], OLS works best when there is a linear trend in the data.

ACKNOWLEDGEMENTS

This study was supported by Universiti Teknologi Malaysia and the Ministry of Higher Education (MOHE), Malaysia.

REFERENCES

1. Siti Normah Hassan, M. H. Ahmad, Suhartono and N. Mohamed, A Comparison of the Forecast Performance of Double Seasonal ARIMA and Double Seasonal ARFIMA Models of Electricity Load Demand, Applied Mathematical Sciences, 6 (135), 2012, 6705–6712.
2. W. S. W. William, Time Series Analysis: Univariate and Multivariate Methods, Pearson Education, USA, 2006.

3. R. M. Sakia, The Box-Cox Transformation Technique: A Review, The Statistician, 41, 1992, 169–178.
4. G. D. Hutcheson, Ordinary Least Squares Regression. The SAGE, 2011.

CITATION

Nor Hamizah Miswan, Pung Yean Ping, and Maizah Hura Ahmad, On Parameter Estimation for Malaysian Gold Prices Modelling and Forecasting, Int. Journal of Math. Analysis, Vol. 7, 2013, no. 22, 1059–1068.

Causality Using Local Measures of Divergence

Mehrdad and Jafari-Mamaghani

Department of Mathematics, Stockholm
University, Stockholm, Sweden
Department of Biosciences & Nutrition,
Karolinska Institutet, Huddinge, Sweden

ABSTRACT

The employment of Granger causality analysis on temporal data is now
a standard routine in many scientific disciplines. Since its inception,
Granger causality has been modelled using a wide variety of analyti-
cal frameworks of which, linear models and derivations thereof have
been the dominant choice. Nevertheless, a body of research on Granger causality and its applications has focused on non-linear and non-
parametric models. One common choice for such models is based on
employment of multivariate density estimators and measures of divergence. However, these models are subject to a number of estimations
and tuning components that have a great impact on the final outcome.
Here we focus on one such general model and improve a number of
its tuning bodies. Crucially, we i) investigate the bandwidth selection
issue in kernel density estimation, and ii) discuss and propose a solution to the sensitivity of estimated information theoretic measures of
divergence to non-linear correspondence. The resulting framework of
analysis is evaluated using varied series of simulations.

INTRODUCTION

Analysis of causality, as a natural extension of correlative analysis, has been of great interest in many scientific disciplines. Causal dynamics can be quantified and studied in as diverse areas as social events behind rising levels of criminality, the effects of protein expression on cellular motility, the causes behind growing concentrations of carbon dioxide in the atmosphere, the driving forces leading to increased inflation, and the effects of drug treatments on neuronal activity. The general concept of causality has a history stretching as far back as that of philosophical thinking. However, quantitative concepts of causality are relatively new and have been subject to much debate. Given the wide variety of views on quantitative definitions of causality, it is of little surprise that different schools of thought on this matter have generated a considerable spectrum of quantitative frameworks to define, model and analyze data-driven causal phenomena. Among these, Bayesian networks, differential equation-driven systems analysis and Granger causality have been the dominant frameworks of causal modeling and analysis [17,28]. Here, we address the concept of Granger causality. The concept of Granger causality was formulated in different lights by the works of [18, 48] and consolidated by Clive W.J. Granger in [15]. Granger causality is a particular definition of causality where using temporal resolution, a variable is said to Granger-cause another if the earlier values of the former can enhance the prediction of the present value of the latter in the presence of the latter's earlier values. This particular definition of causality presumes a temporal signal asymmetry where the cause precedes the effect and where the information embedded in the causal variable about the occurrence of the effect conditioned on all other embedded information is unique [15, 22]. Expressed using the mathematical language of probability theory, under H0, given k lags and variables A, B and C, {B} does not Granger-cause {A} at observation index t, if

$$H_0 : A_t \perp \{B_{t-1}, ..., B_{t-1}\} | \{A_{t-1}, ..., A_{t-k}, ...C_{t-k}\}$$

(1)

The statement above can be tested by comparing the two conditional probability densities (CPDs) below [12]:

$$f\left(A_t \,\middle|\, A_{t-1}, ..., A_{t-k}, C_{t-1}, ..., C_{t-k}\right)$$

(2)

$$f\left(A_t \,\middle|\, A_{t-1}, ..., A_{t-k}, B_{t-1}, ..., B_{t-k}, C_{t-1}, ..., C_{t-k}\right)$$

(3)

Under H_0, where {B} does not Granger-cause {A}, the CPDs (2) and (3) are identical. For the sake of convenience let us implement the following substitutions: $X = A_t$, $Y = \{B\}_{t-1}^{t-k}$ and $Z = \left\{\{A\}_{t-1}^{t-k} \{C\}_{t-1}^{t-k}\right\}$.

Thus, it is understood that all formulations, estimations and tests, regardless of their appearance are easily identifiable with any arbitrary multivariate setting. The conditional density functions in (2) and (3) are thus expressed more economically as $f_{X|Z}$ and $f_{X|Y\,Z}$, respectively. Accordingly, (1) can be redefined as:

$$H_0 : X \perp Y|Z$$

(4)

Since the introduction of the concept of Granger causality in 1969, for which Granger received the Nobel Memorial Prize in Economics in 2003, this particular definition of causality has led to a wide array of research focused on its theoretical formulation, data-driven analyses, and expansions of practical routines using various statistical tools. Models of Granger causality have extensively been reviewed in [22, 30, 36, 40]. Most prominently, Granger causality has had a long range of applications within the econometric discipline [11, 14, 15, 39, 43]. Another prominent field of application includes biologically oriented domains such as biological networks inference, and systems modelling and analysis [21, 32, 35, 46].

In this study we will enhance the estimates of multidimensional probability densities used in non-parametric analysis of Granger causality by improved means of bandwidth selection, and address the sensitivity of estimated information theoretic measures to non-linear relationships through the employment of local estimates. The arrangement of the study is as follows. In Methods, we present the elementary models of Granger causality, review the non-parametric extensions using multivariate density estimation and measures of divergence, and address the bandwidth selection and sensitivity issues mentioned above. In Results, we investigate the performance of the improved non-parametric framework using series of simulations. Lastly, in Discussion we draw upon the discussed themes and conclude the study.

METHODS

Parametric Models

Methods of linear regression have been the most frequently employed tools to model and test the presence of Granger causality. Using linear regression, the hypothesis in (1), omitting the variable(s) C for notational convenience, is tested using the models

$$H_0 : A_t = \alpha_0 + \sum_{i=1}^{k} \beta_i A_{t-i} + \varepsilon_t$$

$$H_1 : A_t = \alpha_0 + \sum_{i=1}^{k} \beta_i A_{t-i} + \sum_{i=1}^{k} \gamma_i B_{t-i} + \eta_t$$

where $\alpha.$ are the regression intercept coefficients, $\beta.$ and $\gamma.$ are the regression variable coefficients, and the residual terms ε_t and η_t are independent and identically distributed according to a standard Gaussian $N(0, \sigma^2)$. See [13] for a generalization of the models above.

Among other techniques, the models above can be tested using the Granger-Sargent test [15], also known as the structural Chow test in econometric literature [9]:

$$F = \frac{(RSS_\varepsilon - RSS_\eta)/k}{RSS_\eta/(n-2k)} \sim F(k, n-2k)$$

$$(5)$$

where RSS_ε is the 'restricted' residual sum of squares under H_0, RSS_η is the 'unrestricted' residual sum of squares under H_1, n is the number of observations, and k is the number of included lags.

Naturally, these models assume linear specifications of the functional form between the regressors and the response variable. Additionally, given the widespread usage of ordinary least squares' (OLS) regression, the models demand the fulfilment of the standard assumptions of homoscedasticity, error normality, and lack of multicollinearity. A routine practice to relax the latter assumption is achieved by principal components regression or partial least squares regression at the possible expense of predictive power. A partial circumvention of the other restrictions is outlined in the following. An approach to relax the strict linearity of the models above is to use functional derivations or higher-dimensional projections of variables. Thus, the hypothesis in (1) can be tested using the models

$$H_0 : A_t = \alpha_0' + \sum_{i=1}^{k}\sum_{j=1}^{J_i} \beta_{ij} f_{ij}(A_{t-i}) + \epsilon_t'$$

$$H_1 : A_t = \alpha_1' + \sum_{i=1}^{k}\sum_{j=1}^{J_i} \beta_{ij} f_{ij}(A_{t-i}) + \sum_{i=1}^{k}\sum_{j=1}^{J_i} \gamma_{ij} f_{ij}(B_{t-i}) + \eta_t'$$

where f_{ij} denotes the jth functional derivation of a variable at lag $t - i$, and J_i represents the number of such derivations for each variable at each lag $t-i$. Derivations and projections of this type can naturally be applied to the response variable as well, leading to multivariate models of regression. Typical derivations and projections include polynomial expansions, power transformations, spline-based expansions, and radial basis function expansions. The hypothesis modelled above can be similarly tested using the Granger-Sargent statistic in (5). Specific studies in the modelling and application of non-linear models of regression in the context of Granger causality include [1, 5, 7, 27]. Although offering an enhanced degree of flexibility, models of non-linear regression as formulated above differ from those of linear regression merely in the sense that they model the linear correspondences between derivations of the response and the regressors. Hence, the obtained specifications of the relationships will still be linear or piece-wise linear. Whether or not the modelled relationships are the true underlying patterns of correspondence is not a trivial task to verify. Moreover, higher-dimensional projections of data may need adequate treatments as they often lead to overparametrization and ill-conditioned covariance matrices.

Non-parametric Models

As discussed in the previous section, models of linear regression and the corresponding parametric tests lose their practical validity where assumptions are violated. As a simple example, regression via least squares' techniques with the Granger-Sargent test in (5) perform rather poorly in the presence of variables with skewed distributions. More important however, is the potential exacerbation of performance inadequacy in the case of linear model specification where the true forms of correspondence between the response and the regressors may be non-linear.

Based on these concerns, efforts to improve the modeling of Granger causality have focused on relaxing the assumptions of parametricity in general, and liberation from linear forms of correspondence in particular. Here we will focus on the circumvention of these assumptions by using means of kernel density estimation and non-parametric (information theoretic) measures of divergence. Frameworks of this breed have been reviewed and discussed to varying extents in [22, 36]. In the current study however, our aim is to improve some of the overlooked aspects of non-parametric and information theoretic-driven modeling of Granger causality. These include the practices involved in multivariate kernel density estimation and the sensitivity of information theoretic measures to non-linear relationships.

In the following we outline our framework to test the hypothesis formulated in (4) by comparing the CPDs (2) and (3). Consequently, the procedure of modeling and testing (4) can be divided into three domains: i) estimation of the CPDs, ii) choosing a measure of divergence and, iii) utilization of a suitable framework for tests of significance.

Estimating CPDs: Kernel Density Estimation

Traditionally, histogram density estimators based on equidistant, spline-based or adaptive partitioning of the observation space, have been the preferred solution for estimating probability density functions. Since their introduction however, kernel density estimation (KDE) has become the dominant solution to probability density estimation due to greater flexibility and superior performance [34, 38, 47]. More specifically, when compared to histogram density estimators, the KDE approach leads to better rates of convergence and less sensitivity to binning [41].

Having obtained independent and identically distributed samples $(x_1 \dots x_n)$ from a distribution with an unknown density f, the estimation of f using KDE reduces to the following:

$$\hat{f}_h(X) = \frac{1}{nh} \sum_{i=1}^{n} K\left(\frac{X - x_i}{h}\right)$$

Where K $(\cdot)$ is a kernel function and h is a bandwidth, a smoothing parameter. Common kernel functions are the uniform, triangular, Epanechnikov, or the Gaussian [20, 38]. The latter function, the standard Gaussian kernel $K(x)\phi(x) = e^{-x^2/2}\sqrt{2\pi}$ will be our choice throughout this study.

Thus, using KDE, $f_{X|Z}$ and $f_{X|YZ}$ are estimated according to the following:

$$\hat{f}_{X|Z} = \frac{\hat{f}_{XZ}}{\hat{f}_Z} = \frac{C_{XZ}^{-1}\left[\sum_{i=1}^{n} K\left(\frac{X - x_i}{h^{(X)}}\right) K\left(\frac{Z - z_i}{h^{(Z)}}\right)\right]}{C_Z^{-1}\left[\sum_{i=1}^{n} K\left(\frac{Z - z_i}{h^{(Z)}}\right)\right]}$$

(6)

$$\hat{f}_{X|YZ} = \frac{\hat{f}_{XYZ}}{\hat{f}_{YZ}} = \frac{C_{XYZ}^{-1} \left[\sum_{i=1}^{n} K\left(\frac{X-x_i}{h^{(X)}}\right) K\left(\frac{Y-y_i}{h^{(Y)}}\right) K\left(\frac{Z-z_i}{h^{(Z)}}\right) \right]}{C_{YZ}^{-1} \left[\sum_{i=1}^{n} K\left(\frac{Y-y_i}{h^{(Y)}}\right) K\left(\frac{Z-z_i}{h^{(Z)}}\right) \right]}$$

(7)

Where C. are the normalizing constants, i.e

$$C_{XYZ} = N \left(\prod_{j=1}^{d_X} h_j^{(X)} \right) \left(\prod_{j=1}^{d_Y} h_j^{(Y)} \right) \left(\prod_{j=1}^{d_Z} h_j^{(Z)} \right)$$

(8)

Where $N = n_{dX} + n_{dY} + n_{dZ}$ is the total number of observations used in the estimation, d_X is the number of dimensions of X, and $h^{(X)}$. Are the bandwidths associated with the corresponding dimensions of X; resulting in an individual bandwidth for every lag of every variable in the KDE of the CPDs. The other definitions follow accordingly.

Varieties of KDE in non-parametric modelling of Granger causality have been previously employed to test the hypothesis in (4) in a number of studies [8, 21, 35]. Specifically, the kernel function chosen in [8] is $K(x) = (3 - x^2)\ \phi(x)/2$ and the bandwidths are evaluated using a 'rule of thumb'. Notably, the choice of bandwidths is an issue that demands a discussion.

Bandwidth Selection

Adequate bandwidth selection in kernel density estimation of multi-variate distributions is critical to obtaining reliable estimations. The bandwidth, as a smoothing parameter, determines the number of observations included in 'windows' in the process of KDE. Specification of the bandwidth size is a clear bias-variance tradeoff where relatively small bandwidths lead to high variance and low bias, and relatively large bandwidths to the opposite [20]. Currently, choosing the bandwidth parameter in a wide variety of applications is determined by a 'rule of thumb' or other application-specific evaluations. Unsurprisingly, such frameworks fail to generalize and when used without critical reflection can lead to erroneous and biased estimations. This issue has to an extent been addressed by other more adaptive methods designed to enhance the unbiasedness of KDE [19, 23, 45]. However, none of the approaches qualifies as a purely non-parametric practice yielding satisfactory density estimations regardless of underlying distributions. Here, in the following employments of KDE, bandwidths are chosen using the plug-in bandwidth selection method proposed in [4] based on linear diffusion processes. Using simulated data, the bandwidths produced according to [4] return reliable and accurate results and consistently outperform the traditional bandwidth selection methods (see Figure 1).

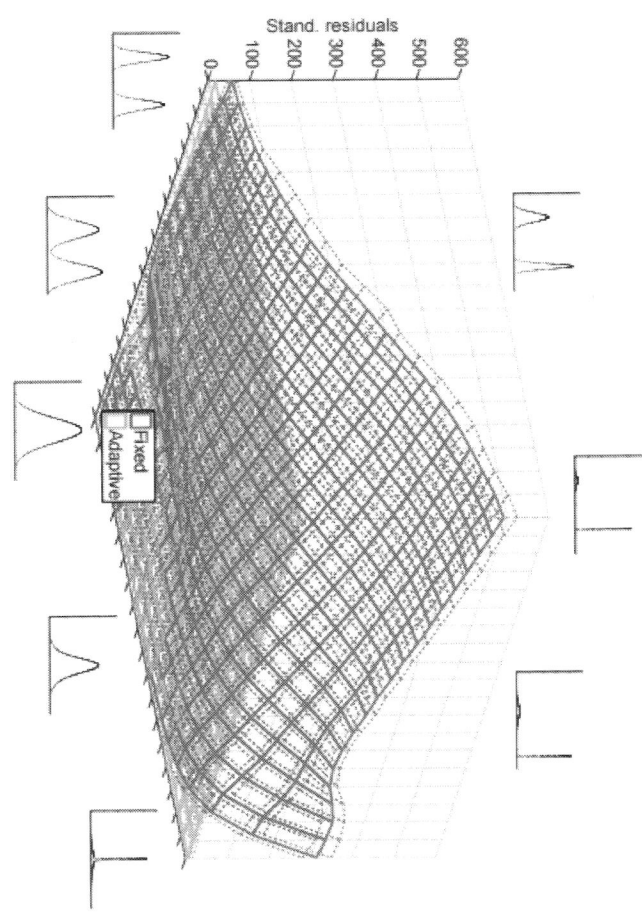

The simulations in Figure 1 are based on two axes of variation: distribution bimodality and mode shape. The source distributions are two-component mixtures of Gaussian distributions with 1000 randomly generated numbers and equal mixing proportions. The bimodality of the distribution is determined by distancing the two Gaussian components and the mode shape is determined by altering the variance of one of the components. Both of these variations are performed progressively for 20·20 bins as seen in Figure 1. The obtained residuals represent the point wise standardized deviations between the estimated densities and the underlying mixture model density. The fixed bandwidth of choice has been $\left(4\hat{\sigma}^5 / 4n\right)^{0.2}$ where $\hat{\sigma}$ the standard deviation of the sample is and n is the number of samples. This bandwidth is also known as Silverman's rule of thumb [38]. As seen in Figure 1, the adaptive bandwidths proposed by [4] clearly outperform the fixed bandwidths in terms of unbiasedness. Thus, in the remainder of this study all bandwidths in all instances of KDE are realized using the method in [4].

Measures of Divergence

As stated in the Introduction, the existence of Granger causality is equivalent to an inequality between the conditional probability densities f_{XZ} and f_{XYZ}. Under the null hypothesis (4) the CPDs above have to be equal. This plausible equality or its lack thereof can be quantified using a variety of measures of divergence. Among others, the weighted Hollinger distance and the Euclidean distance have been used to quantify the divergence between the kernel densities estimated CPDs in [8], of which the latter will be employed in a number of our simulations. More specifically, the Euclidean distance between (7) and (6) is defined as [44]:

$$d_E = n^{-1}\left[\sum_{i,j} ||(f_{X|YZ})_i - (f_{X|Z})_j|| - \right.$$

$$\left. \frac{1}{2}\sum_{i,j}||(f_{X|YZ})_i - (f_{X|YZ})_j|| - \frac{1}{2}\sum_{i,j}||(f_{X|Z})_i - (f_{X|Z})_j||\right]$$

(9)

Where $(f.|.)_i$ is the ith realization of $(f.|.)$, and n is the number of such realizations.

In the remainder the focus will be on measures of divergence derived from the field of information theory. This focus is motivated by the inherent nonparametricity of information theoretic measures of divergence. Firstly, given empirically estimated distribution functions, there is no need for distribution specificity. Secondly, information theoretic measures operate in an evidence based manner allowing detection of non-linear relationships, the practical validity of which will be discussed shortly.

The first discussed measure for quantification of divergence between f_{xyz} and f_{xz} is the Jensen-Shannon divergence introduced in [26]. Given the right weighting parameters, the Jensen-Shannon divergence can be regarded as the symmetrized version of the Kullback-Leibler divergence [25, 26]. However, before a thorough elaboration of these measures, a review on some of the basic concepts in information theory may be useful.

In information theory the (differential) Shannon entropy of a random variable X with a continuous probability distribution f_X with support on x is defined as [10, 37]:

$$H(X) \equiv -E[\log_b f_X] = -\int_x f x \log_b f x dx \qquad (10)$$

Where b is the base of the logarithm determining the terms in which the entropy is measured (e.g. b = 2 for bits and b = e for nats). The higher the value of entropy the higher the uncertainty associated with the outcomes of the studied random variable. The most illustrative demonstrations are those of bernoulli trials such as coin tosses. A symmetric coin $(p_1 = p_2 = 0.5)$ yields the highest entropy value whereas values of entropy decrease with increasing coin asymmetry; hence decreasing uncertainty

The Kullback-Leibler divergence between two probability distribution functions f_X and g_X with support on X is defined as [10, 25]:

$$D_{KL}(f_X||g_X) \equiv \int_X f_X \log_b \frac{f_X}{g_X} dx$$

(11)

Note that replacing the arguments f_X and g_X in (11) can lead to a different quantification of divergence. Therefore, the Kullback-Leibler divergence is not regarded as a symmetric measure of divergence.

Given the definitions above, the Jensen-Shannon divergence between two probability distributions f_X and f_Y representing two stochastic variables X and Y, respectively, with identical support is defined by

$$D_{JS}(f_X||f_Y) \equiv \pi_X D_{KL}(f_X||m_{XY}) + \pi_Y D_{KL}(f_Y||m_{XY})$$

(12)

Where $m_{XY} = [f_X + f_Y]/2$, and π. denotes the weights assigned to each distribution subject to $\pi_X + \pi_Y = 1$. In our applications, unless otherwise stated, $\pi. = 1/d$ where d is the number of dimensions. Moreover, $0 \leq D_{JS}(\cdot||\cdot) \leq \log_b(b)$. Expressed in terms of Shannon entropy, (12) can be redefined as:

$$D_{JS}(f_X||f_Y) \equiv H(\pi_X X + \pi_Y Y) - \pi_X H(X) - \pi_Y H(Y)$$

(13)

Thus, with $\pi = 0.5$, the symmetrized divergence between the estimated CPDs (7) and (6) is evaluated according to:

$$
\begin{aligned}
D_{JS}\left(f_{X|YZ}||f_{X|Z}\right) = & H\left(\pi[X|YZ] + \pi[X|Z]\right) - \pi H\left(X|YZ\right) - \pi H\left(X|Z\right) \\
= & -\int_{XYZ} \left[\pi f_{X|YZ} + \pi f_{X|Z}\right] \log_b \left[\pi f_{X|YZ} + \pi f_{X|Z}\right] dxdydz \\
& + \pi \int_{XYZ} f_{X|YZ} \log_b \left[f_{X|YZ}\right] dxdydz \\
& + \pi \int_{XZ} f_{X|Z} \log_b \left[f_{X|Z}\right] dxdz
\end{aligned}
$$

(14)

Under H_0: $X \perp Y | Z$, it is easily seen that $D_{JS}\left(\hat{f}_{X|YZ}||\hat{f}_{X|Z}\right) = 0$. Another information theoretic measure of divergence frequently used in the context of Granger causality has been the conditional mutual information. In fact, there is an illuminating relationship between (14) and conditional mutual information which is elaborated in the Appendix 5.1.

The conditional mutual information of X and Y given Z, with supports on x , y, and z respectively, can be expressed as:

$$
\begin{aligned}
I(X;Y|Z) &= H(X|Z) + H(Y|Z) - H(XY|Z) \\
&= H(XZ) + H(YZ) - H(XYZ) - H(Z) \\
&= H(X|Z) - H(X|YZ) \\
&= \int_{xyz} f_{XYZ} \log_b \frac{f_{XY|Z}}{f_{X|Z} f_{Y|Z}} dx dy dz
\end{aligned}
\tag{15}
$$

The conditional mutual information is symmetric: I(X; Y |Z) = I(Y; X|Z). Furthermore, I(X; Y |Z) ≥ 0 with equality if and only if X and Y are independent. Lastly, I(X; X|Z) = H (X|Z).

As the conditional entropy of a set of variables could vary, the conditional mutual information in (15) should be normalized to compensate for possible fluctuations. In this study, we will employ the normalized conditional mutual information and conditional symmetric uncertainty:

$$
NI(X;Y|Z) = \frac{I(X;Y|Z)}{\sqrt{H(X|Z)H(Y|Z)}}
\tag{16}
$$

$$
SU(X;Y|Z) = \frac{2.1(X;Y|Z)}{H(X|Z) + H(Y|Z)}
\tag{17}
$$

Other studies outside of the context of Granger causality focused on normalization of mutual information include [6, 42, 49]. It should be noted that a number of information theoretic measures used in the context of Granger causality can be deducted to a derivation of conditional mutual information. One of these measures is the widely used 'transfer entropy' as proposed in [33], which can be reparametrized as the conditional mutual information [22, 36]. Interestingly, the transfer entropy is shown to have a functional relationship with the linear estimators of Granger causality [3,29]. Other measures coined under the term 'directed influence' also fall under this category as derivations of conditional mutual information [36].

Sensitivity to Linearity

Based on our observations and as put forward in [24,31], information theoretic measures estimated via KDE show a quantifiable and non-random sensitivity to non-linear patterns of correspondence. Thus, given the same degree of noise, information theoretic measures estimated via KDE assign lower scores to nonlinear relationships than those of a linear manner. As this contradicts the non-parametric nature of information theoretic measures a series of different approaches have been devised to circumvent this issue. As outlined in [31], limiting the sensitivity of information theoretic measures (e.g. mutual information) to non-linearity may be achieved by means of domain partitioning. However, the partitioning devised in [31] demands a large number of observations and due to its extensive partitioning is limited to the bivariate case. It is understood that in the present case of time-series modelling using two or more variables and one or more lags, the smallest number of dimensions available for partitioning is three. Moreover, as the number of variables and lags increase, the 'curse of dimensionality' will swiftly necessitate the collection of ever larger deposits of observations.

To circumvent this issue, we propose local estimates of the measures above (14), (16) and (17) as follows:

$$D_{JS}^{N}\left(f_{X|YZ}\|f_{X|Z}\right) = \frac{1}{n}\sum_{i=1}^{n}D_{JS}\left(f_{X^{(N_i)}|Y^{(N_i)}Z^{(N_i)}}\|f_{X^{(N_i)}|Z^{(N_i)}}\right)$$

$$(18)$$

$$NI^{\mathcal{N}}(X;Y|Z) = \frac{1}{n}\sum_{i=1}^{n} NI\left(X^{(\mathcal{N}_i)};Y^{(\mathcal{N}_i)}|Z^{(\mathcal{N}_i)}\right)$$

(19)

$$SU^{\mathcal{N}}(X;Y|Z) = \frac{1}{n}\sum_{i=1}^{n} SU\left(X^{(\mathcal{N}_i)};Y^{(\mathcal{N}_i)}|Z^{(\mathcal{N}_i)}\right)$$

(20)

Where n is the number of observations, Ni denotes the neighbourhood of observation i and vectors $X^{(N_i)}$, $Y^{(N_i)}$ and $Z^{(N_i)}$ represent the observation values in neighbourhood N_i. Naturally, the ability of these estimates to capture non-linear relationships depends on the size of the neighbourhood. Adequate neighbourhood sizes should be chosen subject to their ability to capture local relationships. Small neighbourhood sizes would potentially suffer from large variance whereas neighbourhoods of large proportions might be increasingly biased and fail to capture local structures. When applied to time-series modeling, neighbourhood selection is most intuitively determined by basing it on the domain of the response variable (the 'effect' of the 'cause' variables). Simulations (see Figure 8) confirm the ability of these estimates to detect non-linear relationships.

Tests of Significance

Although there are distribution parameterizations for some information theoretic measures under specific conditions [36], given our aim to constitute a non-parametric framework for modelling and testing, we choose to employ bootstrap resampling to create estimations of probability distribution for the chosen measures of divergence under the null hypotheses. The bootstrap resampling is conducted B times with replacement under $H_0 : X \perp Y | Z$ where at each instance of resampling the empirical observations in Y are randomly permuted to artificially create the independence stated under H_0. Comparing the empirically quantified measure of interest with the bootstrapped distribution under H0 yields the relevant p-values. Naturally, depending on the chosen measure, violations of H_0 can occur under the lower, e.g. $D_{JS}(f_{XYZ}\|f_{XZ})$ or above the upper quantiles, e.g. $I(X; Y | Z)$, of the bootstrapped distributions.

RESULTS

The following series of simulations are designed to evaluate the performance of the hitherto discussed framework of Granger causality analysis. In the bivariate and functional time-series, the parametric framework stated in the Methods has also been included. The focus in the functional time-series is on (19) as this measure performs nearly identically as (18) and (20). In the multivariate time-series, the aim is to investigate the ability of the framework in detection of causal links in high-dimensional spaces.

Bivariate Time-series

In the simulated bivariate time-series only one variable is designed to be self-generative. More specifically, we define two variables X and Y, and denote the temporal index by t, $x_t = a$ and $y_t = b$ where a and b are random numbers drawn from the standard Gaussian distribution. The remaining is defined as:

$$\begin{cases} x_{t-i} = a + i \cdot \epsilon & i = 1, \ldots, k \\ y_{t-i} = a + g_c(i) \cdot \epsilon & i = 1, \ldots, k \end{cases}$$

Where $\varepsilon \sim N(0, 1)$, $k = 10$ and $g_c(i) = c \cdot i$ determines the degree to which lags of y are designed to correlate with x_t . Common choices for c in the following simulations include 0.1,1,2,5 and 10. Each realization of X and Y consists of 100 $\{x_t\}$ and $\{y_t\}$, respectively. That is, 1100 sample points in X and Y. At each lag k the following hypotheses are tested:

$$H_0^x : \{X_t\} \perp \{Y_{t-1}, ..., Y_{t-k}\} | \{X_{t-1}, ..., X_{t-k}\}$$
$$H_0^y : \{Y_t\} \perp \{X_{t-1}, ..., X_{t-k}\} | \{Y_{t-1}, ..., Y_{t-k}\}$$

The hypotheses H_0^x and H_0^y are tested via the classical linear regression framework in 2.1. In addition, using density estimation via KDE as devised in (7) and (6), the hypotheses above are tested by employing the Euclidean distance (9), the normalized conditional mutual information (16), the conditional symmetric uncertainty (17), the Jensen-Shannon divergence (12). The results based on 100 realizations of X and Y using B = 1000 in the bootstrap resampling's, are displayed in Figures 2, 3, 4, 5, 6, and summarized in Figure 7. The arrangement of panels in Figures 2-6 is as follows: panels along the horizontal axis represent the noise levels 0.1,1,2,5 and 10 whereas panels along the vertical axis represent the number of included lags: 1, 5 and 10. The gray probability masses represent the reference distributions under the null hypotheses H_0^x . These distribution are either evaluated analytically (as for the Granger-Sargent test), or obtained computationally using the bootstrap resampling scheme outlined above. The crosses on the x-axes of these probability masses represent the 'empirical' scores of the simulations. It is easily seen that increasing noise levels, regardless of the number of included lags, lead to less frequent rejections of the null hypotheses. That is, the detectability of the synthetic causal link between the two variables is degraded as a function of added noise.

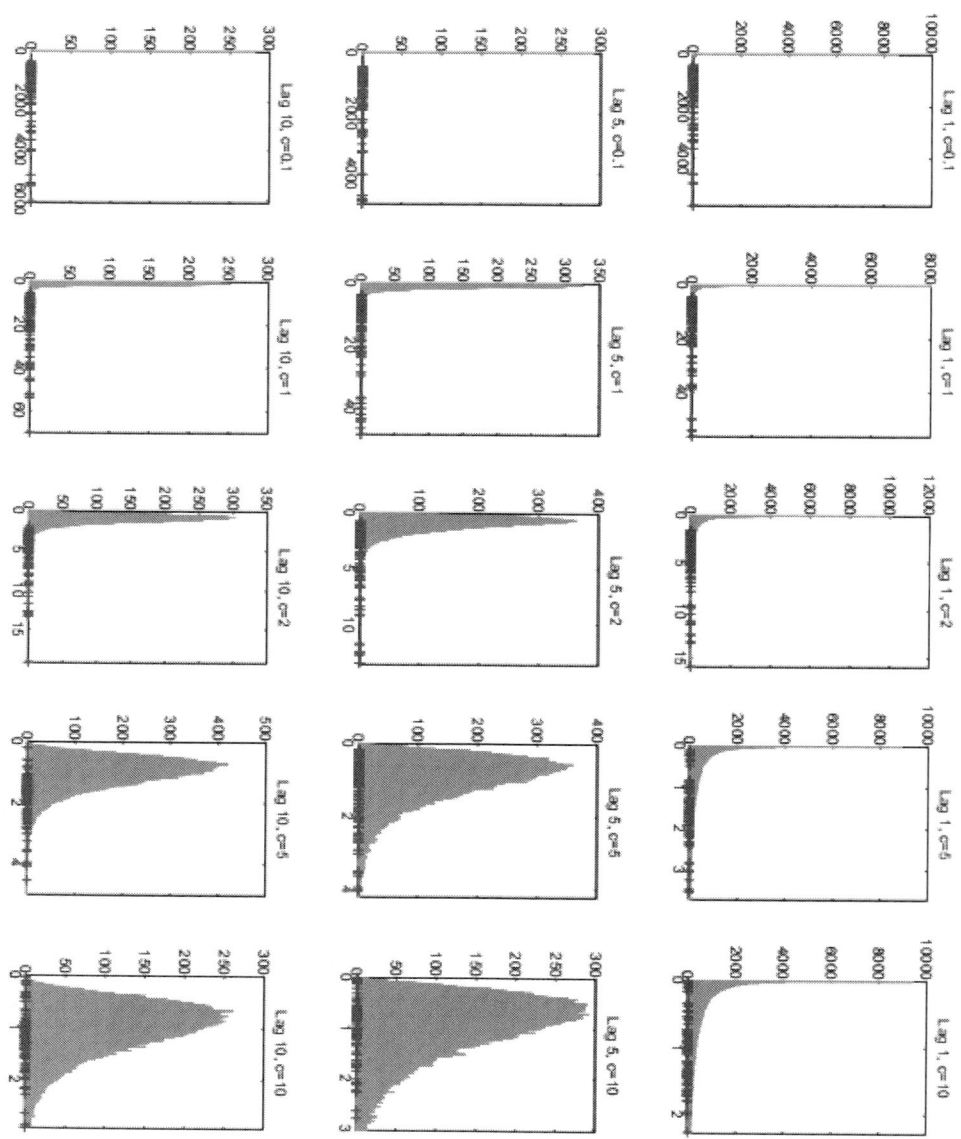

Figure 2: The Granger-Sargent test. Results based on simulations of bivariate time-series for k = 1...10 lags and 5 levels of noise as defined in Bivariate time-series. The + signs on the x-axes represent the yielded scores from the Granger-Sargent test as formulated in (5). The gray probability mass is that of the null hypothesis H_0^x under simulated data using the same test statistic.

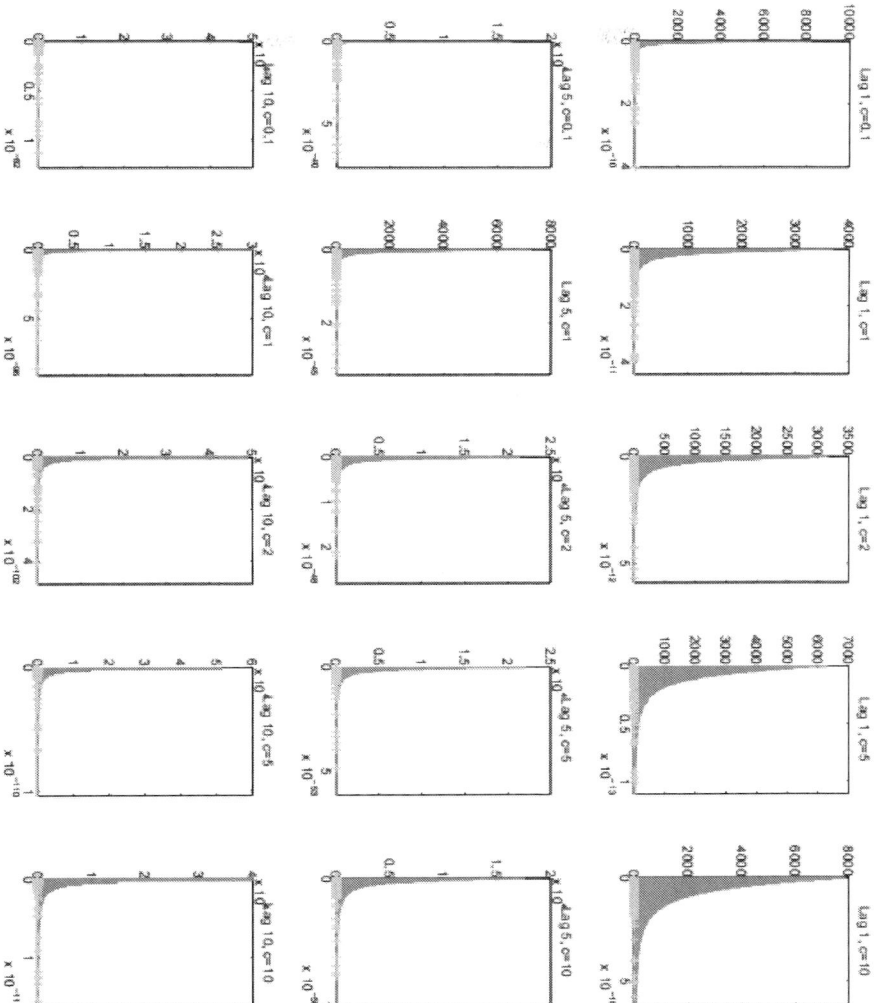

Figure 3: The Euclidean distance. Results based on simulations of bivariate time-series for k = 1...10 lags and 5 levels of noise as defined in Bivariate time-series. The + signs on the x-axes represent the yielded scores from the Euclidean distance as formulated in (9). The gray probability mass is that of the null hypothesis H_0^x under simulated data using the same measure of divergence.

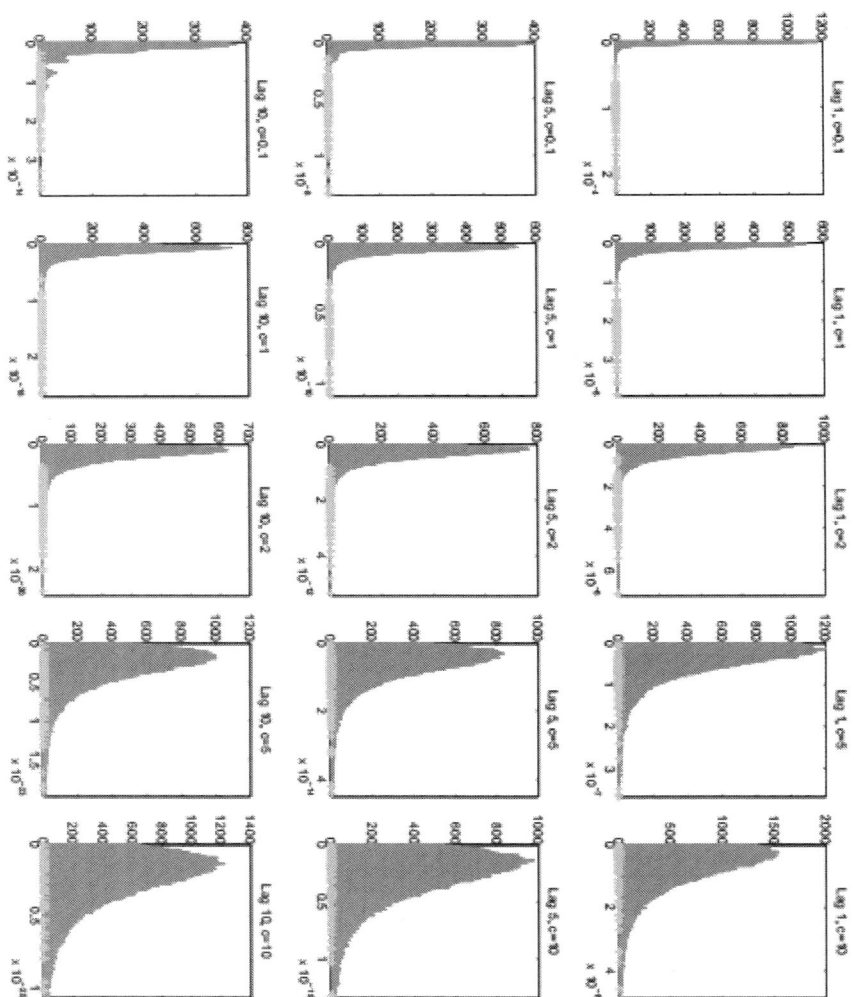

Figure 4: Normalized conditional mutual information. Results based on simulations of bivariate time-series for k = 1…10 lags and 5 levels of noise as defined in Bivariate time-series. The + signs on the x-axes represent the yielded scores from the normalized conditional mutual information as formulated in (16). The gray probability mass is that of the null hypothesis H_0^x under simulated data using the same measure of divergence.

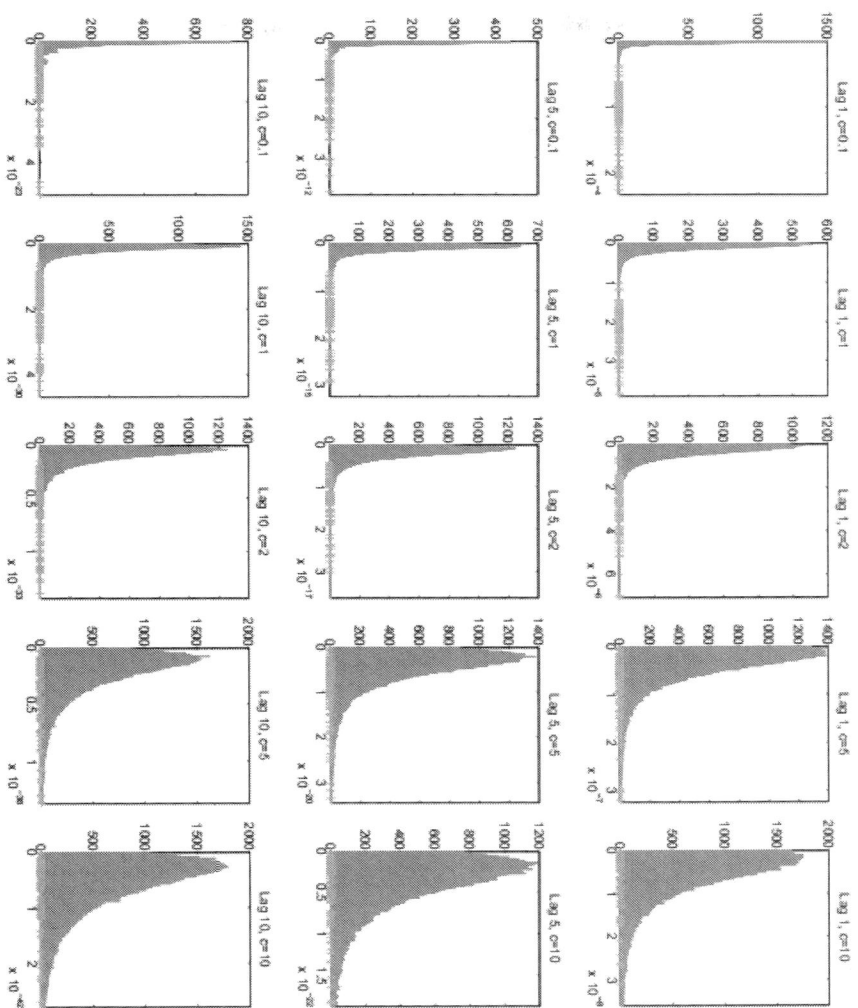

Figure 5: Symmetric uncertainty. Results based on simulations of bivariate time-series for k = 1...10 lags and 5 levels of noise as defined in Bivariate time-series. The + signs on the x-axes represent the yielded scores from the normalized conditional mutual information as formulated in (17). The gray probability mass is that of the null hypothesis H_0^x under simulated data using the same measure of divergence.

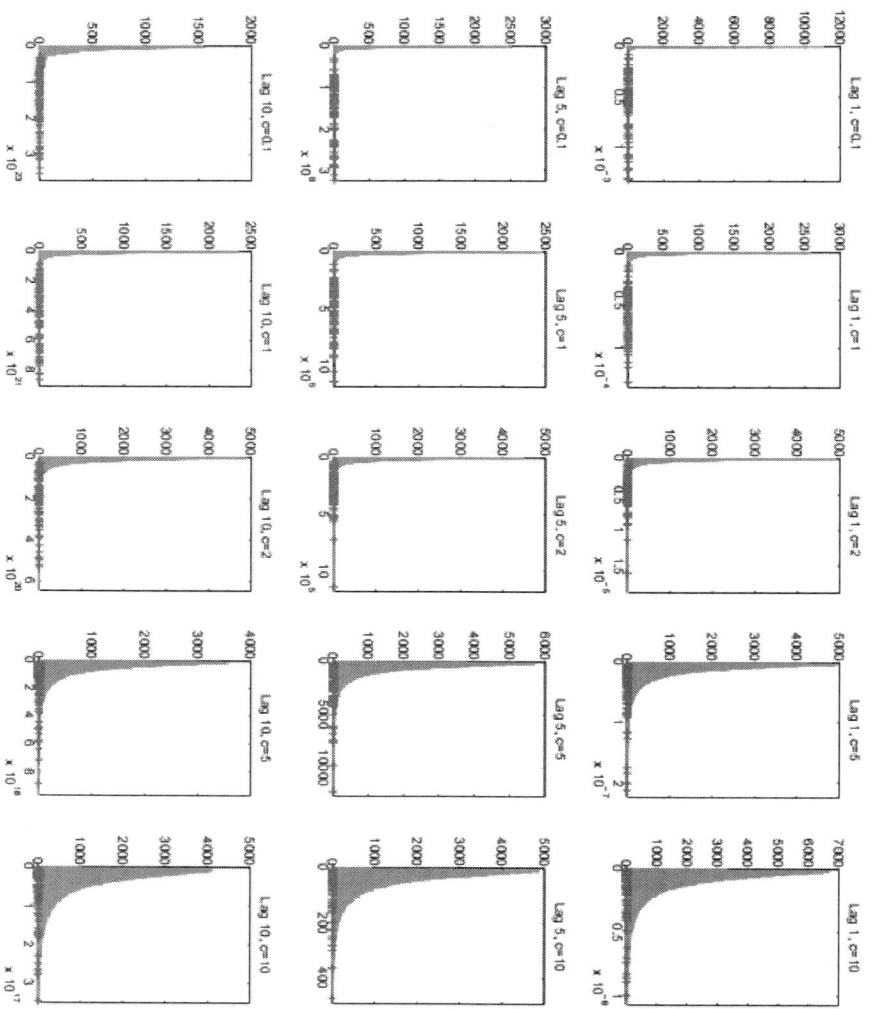

Figure 6: The Jensen-Shannon divergence. Results based on simulations of bivariate time-series for k = 1..10 lags and 5 levels of noise as defined in Bivariate time-series. The + signs on the x-axes represent the yielded scores from the Jensen-Shannon divergence as formulated in (13). The gray probability mass is that of the null hypothesis H_0^x under simulated data using the same measure of divergence.

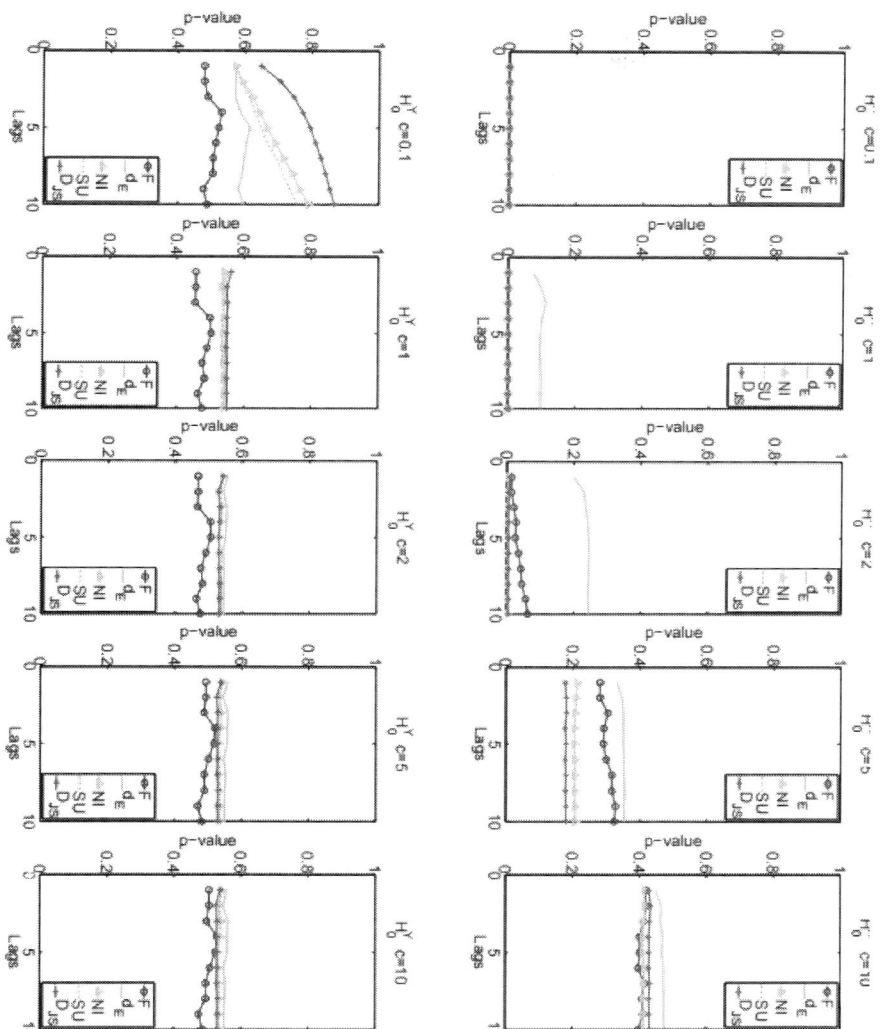

Figure 7: Summary of the bivariate simulations. The aggregate results of the simulated bivariate time-series as defined in Bivariate time-series demonstrated using p-values under the null hypotheses. The five horizontal panels represent the five different noise levels (c) implemented in the simulations. The upper panels represents the results from the hypotheses based on X (H_0^X) being the effect whereas the lower panels represent the results from the hypotheses based on Y (H_0^Y) being the effect. The employed measures of divergence are displayed in the legends.

The summary of the results above under H_0^X and additionally under H_0^Y are represented in Figure 7 in terms of p-values for each noise level at each lag. Here, it is easily seen that the Euclidean distance d_E is outperformed by the Granger-Sargent test and the divergence measures under H_0^X. Additionally, the information theoretic measures of divergence outperform the Granger Sargent test in the detection of causal relationships. Among the information theoretic measures, the Jensen-Shannon divergence performs most optimally.

Functional time-series Here we define two variables X and Y , and denote the temporal index by t, $x_t = a$ and $y_t = b$ where a is a random number generated according to a $U(-2, 2)$ distribution and b is a random number drawn from the standard Gaussian distribution. The remaining is defined as:

$$\begin{cases} x_{t-1} = a + g_c(i).\varepsilon & i = 1,...,k \\ y_{t-i} = \sin(a) + g_c(i).\varepsilon & i = 1,...,k \end{cases}$$

Where $\varepsilon \sim N(0, 1)$, $k = 5$ and $g_c(i) = c \cdot i$ determines the degree to which lags of x and y are designed to correlate with x_t. In this series c is set to three different values: 1, 2.5 and 5. Each realization of X and Y consists of 50 $\{x_t\}$ and $\{y_t\}$, respectively. The null hypothesis is identical to that of the bivariate time-series, tested using the Granger-Sargent test and the neighbourhood-based measure of divergence is normalized conditional mutual information as defined in (19) where the neighbourhood size is chosen to include the 10 nearest neighbors to every observation. The results based on 100 simulations of X and Y using B = 500 in the bootstrap resamplings, are displayed in Figure 8. The groups of scatterplots on the left side of Figure 8 represent the bivariate distributions of X at lag 0 against X and Y at lags k = 1... 5. The scatterplots are arranged vertically top down with increasing noise levels. The three panels on the right side in Figure 8 represent the yielded p-values from the simulations. Regardless of the noise level, the Granger-Sargent test fails to capture the functional causal link between X and Y whereas the neighbourhood-based normalized conditional mutual information NI^N detects the relationship at all noise levels.

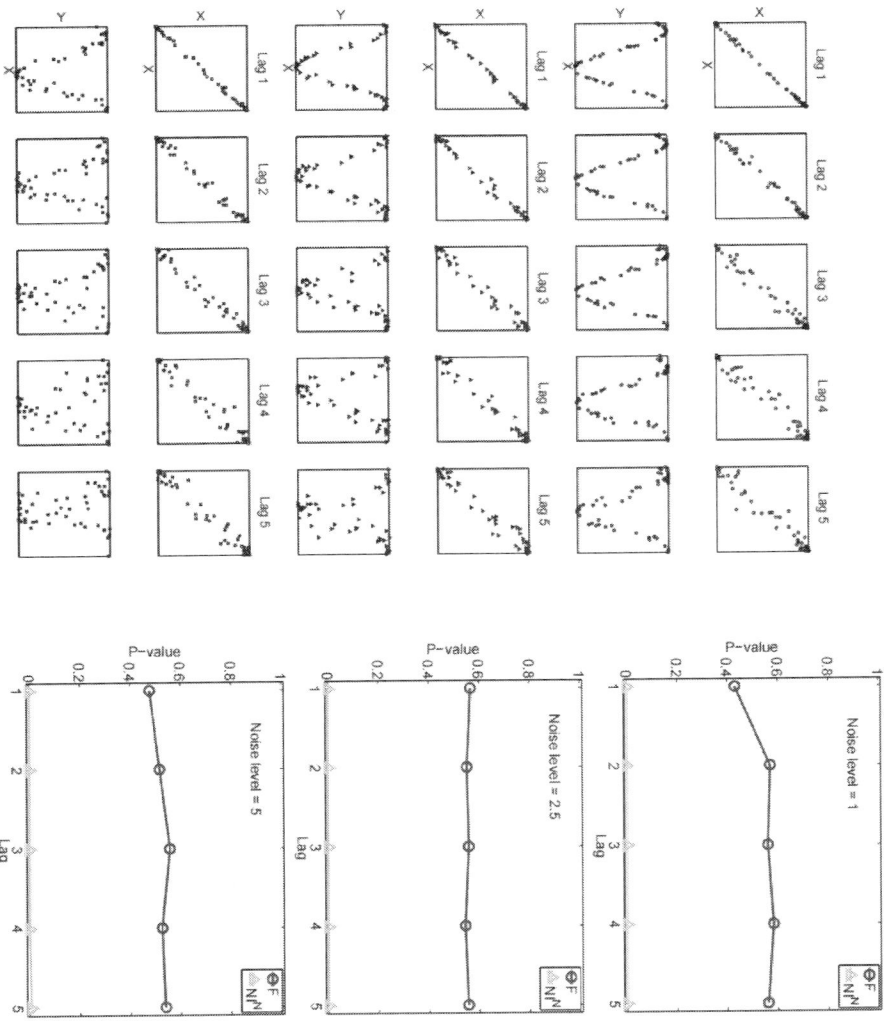

Figure 8: Functional time-series and the neighbourhood-based divergence. The results of the functional time-series simulations as outlined in Functional time-series. The bivariate scatter plots demonstrate the functional relationship between the two variables at each lag and for each noise level. The plotted p-values are associated with the Granger-Sargent test and the neighbourhood-based normalized conditional mutual information.

Multivariate time-series

The simulated multivariate time-series consists of four variables (nodes) W, X, Y, and Z and resembles a type of reverse design in its construction compared to the former bivariate cases.

More specifically, $w_t = a$, $x_t = b$, $y_t = c$ and $z_t = d$, and where a, b, c and d are generated randomly according to the standard Gaussian distribution. Additionally:

$$\begin{cases} w_{t-k} = w_t + g_3(1) \cdot \epsilon \\ x_{t-k} = w_t + g_1(1) \cdot \epsilon \\ y_{t-k} = x_t + g_{1.5}(1) \cdot \epsilon \\ z_{t-k} = x_t + g_{0.15}(1) \cdot \epsilon + y_t + g_{0.3}(1) \cdot \epsilon \end{cases}$$

Where k = 10 and where similarly, each ε is generated randomly according to a N(0, 1). The remaining is constructed according to:

$$\begin{cases} w_{t-i} = w_{t-i-1} + g_3(k - i) \cdot \epsilon & i = 1, \ldots, k - 1 \\ x_{t-i} = w_{t-i-1} + g_1(k - i) \cdot \epsilon & i = 1, \ldots, k - 1 \\ y_{t-i} = x_{t-i-1} + g_{1.5}(k - i) \cdot \epsilon & i = 1, \ldots, k - 1 \\ z_{t-i} = z_{t-i-1} + g_{0.5}(k - i) \cdot \epsilon & i = 1, \ldots, k - 1 \end{cases}$$

This specific setting is designed to test whether the employed framework of KDE, divergence quantification and bootstrap tests of significance can capture time-resolved correlations in high dimensions. The cross-temporal correlations increase as more lags from the 'past' are included in the model. Consequently, during this process the space in which the data is embedded is inflated as more dimensions (lags) are added to the KDE.

The results of the simulation using B = 1000 bootstrap resampling's presented in Figure 9 and 10, based on the Jensen-Shannon divergence and normalized conditional mutual information respectively, will help to illuminate the motivation behind the specific design.

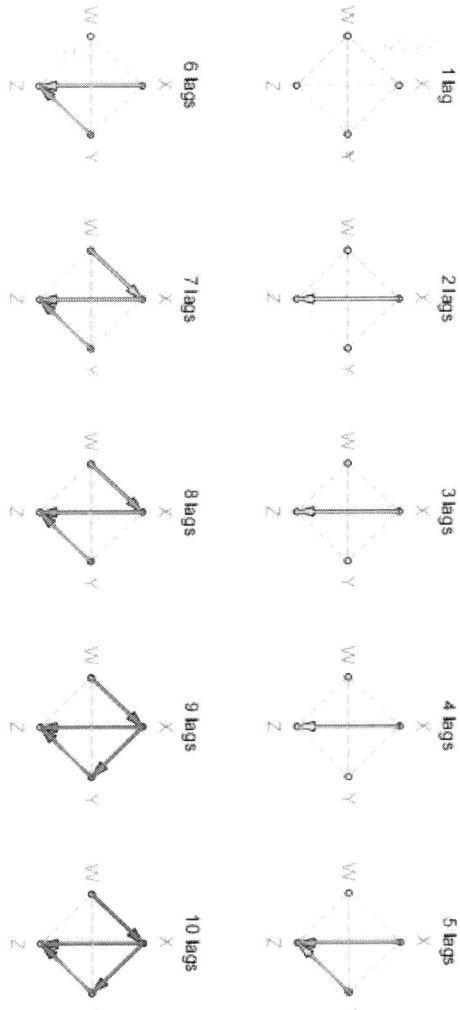

Figure 9: Multivariate time-series results using the Jensen-Shannon Divergence. The aggregate results of the simulated multivariate time-series as outline in Multivariate time-series. The measure of interest here is the Jensen-Shannon divergence. The results are represented in a lag-wise manner as each lag stretching further back into the 'past' reveals a stronger causal relationship on aggregate. The color-coding of the arrows is based on the spearman correlation coefficient of the most recently included lag with the response variable/the effect.

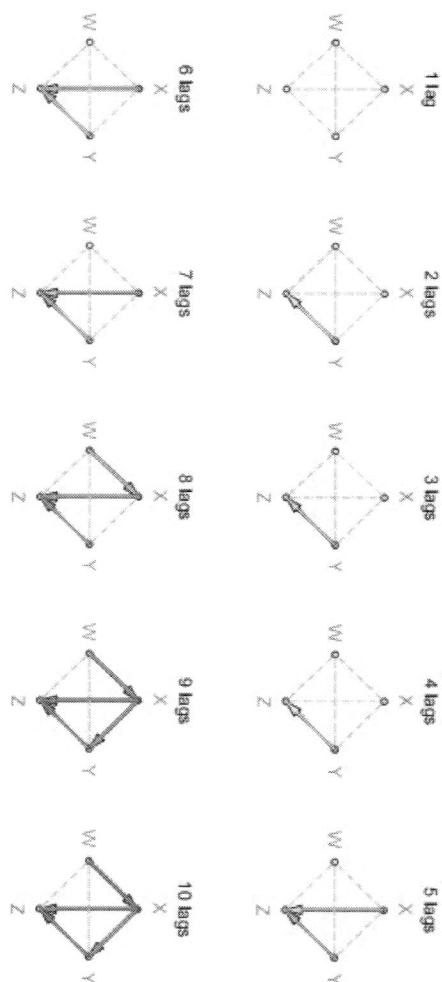

Figure 10: Multivariate time-series results using the normalized conditional mutual information. The aggregate results of the simulated multivariate time-series as outline in Multivariate time-series. The measure of interest here is the normalized conditional mutual information. The results are represented in a lag-wise manner as each lag stretching further back into the 'past' reveals a stronger causal relationship on aggregate. The color-coding of the arrows is based on the spearman correlation coefficient of the most recently included lag with the response variable/the effect.

The demonstration of the simulation results in Figures 9 and 10 is arranged according to the following. The four nodes and their conditional causal links are represented for each progressive inclusion of lags. Any presence of a conditional causal link between any two nodes is marked by a directed arrow (for p-values < 0.05). Absences of conditional causal links are marked by dashed lines. The color-coding of the dashed lines denotes the rank correlation coefficients between one node at lag 0 and the other node at the most recently included lag. The color-coding of the arrows denotes the same quantity between the causes at the most recently included lag and the effect at lag 0. As evident in Figures 9 and 10, the two frameworks perform equally well in unveiling the conditional causal relationships between the nodes. One slight difference however, is the order in which the causal links are detected. Closer investigations revealed that this phenomenon had its roots in the between simulation differences in random number generation. As the causal signals have been embedded in 'late' lags, easing their detectability by looking 'farther back' in time, one significant outcome of these simulations is the power of information theoretic measures to detect conditional causal links in relatively high-dimensional spaces.

DISCUSSION

Given the abundance of multivariate time-series data in social and life sciences, there is an evident and inherent interest in moving beyond stationary correlative analysis to dynamic analysis of time-resolved causal relationships. The concept of Granger causality, although not synonymous with causality itself, offers a powerful framework for analysis of causality in time-series data. Regarding the elaborated methods to model Granger causality, techniques based on assumptions of parametricity (e.g. ordinary linear regression) are superior to non-parametric designs due to their lower computational demands and more accessible interpretability. Nevertheless, meeting the assumptions of parametric models may not always be feasible.

The framework outlined here in this study based on kernel density estimation (KDE) via adaptive bandwidths, information-theoretic measures of divergence, and bootstrap tests of significance, constitutes a fully non-parametric platform for analysis of Granger causality. Additionally, the inferior sensitivity of information theoretic measures derived via KDE to non-linear relationships has been solved using local neighbourhood-based estimates of information theoretic measures.

Furthermore, the results from our extensive simulations, based on synthetic linear and non-linear (functional) causal relationships, confirm the ability of the discussed platform in detecting causal relationships subject to a varying array of signal to noise ratios. In conclusion, both frameworks succeed to reveal the underlying conditional causal relationships between the nodes, matching the engineering of the simulations. On prospective directions of research, unbiased estimates of differential information theoretic measures of divergence should be assigned top priority. As in earlier studies, we have here seen that differential information theoretic measures of divergence estimated via KDE fail to adhere to their non-parametric premise in the presence of non-linear relationships. Possible solutions to this problem are discretization of the data space, altered kernel functions in the density estimates, or higher-dimensional projections of the data space combined with supervised regularizations. Regardless of possible prospective routes, any relevance of further improvements should be judged upon the specific aim and question of the analysis. Overall, we have shown further advances in non-parametric analysis of Granger causality, allowing such type of temporal data analysis to better take advantage of available information given the abundance of non-parametric data in numerous scientific fields such as econometrics, biology and climatology. The choice of non-parametric analysis of Granger causality is further motivated by the ever growing computational power, facilitating considerable increase in the efficiency of such frameworks of analysis.

ACKNOWLEDGEMENTS

The author wishes to thank Joanna Tyrcha and John G. Lock for insightful discussions and feedback. The research leading to these results has received funding from the European Union's Seventh Framework Programme (FP7/2007-2013) under grant agreement # 258068; EU-FP7-

Systems Microscopy NoE; from the Swedish Research Council grant # 340-2012-6011; and from the Centre for Biosciences at Karolinska Institutet.

APPENDIX

Jensen-Shannon Divergence and Conditional Mutual Information

Suppose $X|Z \sim m_{X^Y Z}$ where $m_{XYZ} = \left[f_X|_{YZ} + f_X|_Z \right] / 2$ is a mixture distribution. Furthermore, regard Y as a binary variable where $Y = 0$ if $X|Z \sim f_{X|Z}$ and $Y = 1$ if $X|Z \sim f_{X|Y}Z$ and $\pi = 0.5$. Then $D_{JS}\left(\hat{f}_{XYZ} \| \hat{f}_{XZ}\right)$ as defined in (12) can be derived further as: D

$$
\begin{aligned}
D_{JS}\left(\hat{f}_{X|YZ}\|\hat{f}_{X|Z}\right) &= \pi D_{KL}(f_{X|YZ}\|m_{XYZ}) + \pi D_{KL}(f_{X|Z}\|m_{XYZ}) \\
&= \pi \int_{xyz} f_{X|YZ} \log_b \left[\frac{f_{X|YZ}}{m_{XYZ}}\right] dxdydz + \pi \int_{xyz} f_{X|Z} \log_b \left[\frac{f_{X|Z}}{m_{XYZ}}\right] dxdydz \\
&= \pi \int_{xyz} f_{X|YZ} \left[\log_b f_{X|YZ} - \log_b m_{XYZ}\right] dxdydz \\
&\quad + \pi \int_{xyz} f_{X|Z} \left[\log_b f_{X|Z} - \log_b m_{XYZ}\right] dxdydz \\
&= -\pi \int_{xyz} f_{X|YZ} \log_b m_{XYZ} dxdydz - \pi \int_{xz} f_{X|Z} \log_b m_{XYZ} dxdydz
\end{aligned}
$$

$$
+ \pi \int_{xyz} f_{X|YZ} \left[\log_b f_{X|YZ}\right] dxdydz + \int_{xz} f_{X|Z} \left[\log_b f_{X|Z}\right] dxdydz
$$

$$
= -\pi \int_{xyz} m_{XYZ} \log_b m_{XYZ} dxdydz
$$

$$
+ \pi \int_{xyz} f_{X|YZ} \left[\log_b f_{X|YZ}\right] dxdydz + \int_{xz} f_{X|Z} \left[\log_b f_{X|Z}\right] dxdydz
$$

$$
= H(X|Z) - H(X|YZ) = I(X;Y|Z)
$$

Where I(X; Y |Z) is referred to as the conditional mutual information of X and Y given Z [10]. Similar proofs without using conditional densities can be found in [2, 16, 49].

REFERENCES

1. Nicola Ancona, Daniele Marinazzo, and Sebastiano Stramaglia. Radial basis function approach to nonlinear granger causality of time series. Phys. Rev. E, 70:056221, Nov 2004.

2. Ziv Bar-Yossef. Sampling lower bounds via information theory. In STOC '03: Proceedings of the thirty-fifth annual ACM symposium on Theory of computing, pages 335–344. ACM Press, 2003.

3. Lionel Barnett, Adam B. Barrett, and Anil K. Seth. Granger causality and transfer entropy are equivalent for gaussian variables. Phys. Rev. Lett., 103:238701, Dec 2009.

4. ZI Botev, JF Grotowski, and DP Kroese. Kernel density estimation via diffusion. The Annals of Statistics, 38(5):2916–2957, 2010.

5. David S Broomhead and David Lowe. Radial basis functions, multivariable functional interpolation and adaptive networks. Technical report, DTIC Document, 1988.

6. Nathan D. Cahill. Normalized measures of mutual information with general definitions of entropy for multimodal image registration. In Proceedings of the 4th international conference on Biomedical image registration, WBIR'10, pages 258–268, Berlin, Heidelberg, 2010. Springer-Verlag.

7. Y. Chen, G. Rangarajan, J. Feng, and M. Ding. Analyzing multiple nonlinear time series with extended Granger causality. Phys. Lett. A, 324:26–35, 2004.

8. Nadine Chlaß and Alessio Moneta. Can graphical causal inference be extended to nonlinear settings? In Mauricio Su´arez, Mauro Dorato, and Mikl´os R´edei, editors, EPSA Epistemology and Methodology of Science, pages 63–72. Springer Netherlands, 2009.

9. Gregory C. Chow. Tests of Equality Between Sets of Coefficients in Two Linear Regressions. Econometrica, 28(3):591–605, 1960.

10. Thomas M. Cover and Joy A. Thomas. Elements of Information Theory 2nd Edition. Wiley-Interscience, 2 edition, July 2006.

11. Michael Eichler. Granger causality and path diagrams for multivariate time series. Journal of Econometrics, 137(2):334–353, April 2007.

12. Jean-Pierre Florens and M Mouchart. A note on noncausality. Econometrica: Journal of the Econometric Society, pages 583–591, 1982.

13. John Geweke. Measurement of linear dependence and feedback between multiple time series. Journal of the American Statistical Association, 77(378):304, 1982.

14. John Geweke. Inference and causality in economic time series models. In Z. Griliches and M. D. Intriligator, editors, Handbook of Econometrics, volume 2 of Handbook of Econometrics, chapter 19, pages 1101–1144. Elsevier, September 1984.

15. Clive W. J. Granger. Investigating causal relations by econometric models and cross-spectral methods. Econometrica, 37(3):424–38, 1969.

16. Ivo Grosse, Pedro B. Galv´an, Pedro Carpena, Ram´on R. Rold´an, Jose Oliver, and H. Eugene Stanley. Analysis of symbolic sequences using the Jensen-Shannon divergence. Physical Review E, 65(4):041905+ March 2002.

17. Shuixia Guo, Christophe Ladroue, and Jianfeng Feng. Granger causality: Theory and applications. In Jianfeng Feng, Wenjiang Fu, and Fengzhu Sun, editors, Frontiers in Computational and Systems Biology, volume 15 of Computational Biology, pages 83–111. Springer London, 2010.

18. Trygve Haavelmo. The probability approach in econometrics. Econometrica: Journal of the Econometric Society, 1944.

19. Peter Hall, Tien Chung Hu, and James Stephen Marron. Improved variable window kernel estimates of probability densities. The Annals of Statistics, 23(1):1–10, 1995.

20. Trevor Hastie, Robert Tibshirani, and Jerome Friedman. The Elements of Statistical Learning: Data Mining, Inference, and Prediction. Springer, corrected edition, August 2009.

21. H. Hinrichs, H.J. Heinze, and M.A. Schoenfeld. Causal visual interactions as revealed by an information theoretic measure and fmri. NeuroImage, 31(3):1051–1060, 2006.

22. Katerina Hlav´a˘ckov´a-Schindler, Milan Palu˘s, Martin Vejmelka, and Joydeep Bhattacharya. Causality detection based on information-theoretic approaches in time series analysis. Physics Reports, 441(1):1–46, 2007.

23. M. Jones, I. McKay, and T. Hu. Variable location and scale kernel density estimation. Annals of the Institute of Statistical Mathematics, 46(3):521–535, 1994.

24. Alexander Kraskov, Harald St¨ogbauer, and Peter Grassberger. Estimating mutual information. Physical review. E, 69(6 Pt 2), June 2004.

25. S. Kullback and R. A. Leibler. On Information and Sufficiency. The Annals of Mathematical Statistics, 22(1):79–86, 1951.

26. J. Lin. Divergence measures based on the Shannon entropy. IEEE Transactions on Information Theory, 37(1):145–151, January 1991.

27. Daniele Marinazzo, Mario Pellicoro, and Sebastiano Stramaglia. Nonlinear parametric model for granger causality of time series. Phys. Rev. E, 73:066216, Jun 2006.

28. Judea Pearl. Causality: models, reasoning, and inference. Cambridge University Press, New York, NY, USA, 2000.

29. Mario Pellicoro and Sebastiano Stramaglia. Granger causality and the inverse ising problem. Physica A, 389(21):4747–4754, 2010.

30. Florin Popescu. Robust statistics for describing causality in multivariate time series. Journal of Machine Learning Research - Proceedings Track, pages 30–64, 2011.

31. David N. Reshef, Yakir A. Reshef, Hilary K. Finucane, Sharon R. Grossman, Gilean McVean, Peter J. Turnbaugh, Eric S. Lander, Michael Mitzenmacher, and Pardis C. Sabeti. Detecting Novel Associations in Large Data Sets. Science, 334(6062):1518–1524, dec 2011.

32. R. Salvador, A. Mart´ınez, E. Pomarol-Clotet, S. Sarr´o, J. Suckling, and E. Bull-more. Frequency based mutual information measures between clusters of brain regions in functional magnetic resonance imaging. NeuroImage, 35(1):83–88, March 2007.

33. Thomas Schreiber. Measuring information transfer. Phys. Rev. Lett., 85:461–464, Jul 2000.

34. David W. Scott. Multivariate Density Estimation: Theory, Practice, and Visualization (Wiley Series in Probability and Statistics). Wiley, 1 edition, September 1992.

35. A.-K. Seghouane. Quantifying information flowin fmri using the kullbakcleibler divergence. In Biomedical Imaging: From Nano to Macro, 2011 IEEE International Symposium on, pages 1569–1572, 30 2011-april 2 2011.

36. Abd-Krim K. Seghouane and Shun-Ichi Amari. Identification of directed influence: granger causality, kullback-leibler divergence, and complexity. Neural computation, 24(7):1722–1739, July 2012.

37. Claude E. Shannon. A mathematical theory of communication. Bell System Technical Journal, 27:379–423, July 1948.

38. B. W. Silverman. Density Estimation for Statistics and Data Analysis (Chapman & Hall/CRC Monographs on Statistics & Applied Probability). Chapman and Hall/CRC, 1 edition, April 1986.

39. Christopher A. Sims. Money, Income, and Causality. Amer Econ Rev, 62(4):540–552, 1972.

40. Klaas E. Stephan and Alard Roebroeck. A short history of causal modeling of fMRI data. NeuroImage, January 2012.

41. R. Steuer, J. Kurths, C. O. Daub, J. Weise, and J. Selbig. The mutual information: detecting and evaluating dependencies between variables. Bioinformatics (Oxford, England), 18 Suppl 2(suppl 2):S231–S240, October 2002.

42. Alexander Strehl and Joydeep Ghosh. Cluster ensembles — a knowledge reuse framework for combining multiple partitions. J. Mach. Learn. Res., 3:583–617, March 2003.

43. Liangjun Su and Halbert White. A nonparametric hellinger metric test for conditional independence. Econometric Theory, 24(04):829–864, August 2008.

44. G´abor J Sz´ekely and Maria L Rizzo. Testing for equal distributions in high dimension. InterStat, 5, 2004.

45. George R. Terrell and David W. Scott. Variable Kernel Density Estimation. The Annals of Statistics, 20(3):1236–1265, 1992.

46. P. F. Verdes. Assessing causality from multivariate time series. Phys. Rev. E, 72:026222, Aug 2005.

47. M. P. Wand and M. C. Jones. Kernel Smoothing (Chapman & Hall/CRC Monographs on Statistics & Applied Probability). Chapman and Hall/CRC, 1 edition, December 1994.

48. N. Wiener and P. Masani. The prediction theory of multivariate stochastic processes. Acta mathematica, 98:111–150, 1957.

49. Y. Y. Yao. Information-theoretic measures for knowledge discovery and data mining. In Karmeshu, editor, Entropy Measures, Maximum Entropy Principle and Emerging Applications, pages 115–136, Berlin, 2003. Springer.

CITATION

Mehrdad Jafari-Mamaghani, Non-parametric analysis of Granger causality using local measures of divergence, doi.org/10.12988/ams.2013.35275

Stock Price by Jump Stochastic Time Effective Neural Network Model

Jun Wang, Huopo Pan, and Fajiang Liu
Department of Mathematics, Key Laboratory of Communication and Information System, Beijing Jiaotong University, Beijing 100044, China

8

ABSTRACT

The interacting impact between the crude oil prices and the stock market indices in China is investigated in the present paper, and the corresponding statistical behaviors are also analyzed. The database is based on the crude oil prices of Daqing and Shengli in the 7-year period from January 2003 to December 2009 and also on the indices of SHCI, SZCI, SZPI, and SINOPEC with the same time period. A jump stochastic time effective neural network model is introduced and applied to forecast the fluctuations of the time series for the crude oil prices and the stock indices, and we study the corresponding statistical properties by comparison. The experiment analysis shows that when the price fluctuation is small, the predictive values are close to the actual values, and when the price fluctuation is large, the predictive values deviate from the actual values to some degree. Moreover, the correlation properties are studied by the detrended fluctuation analysis, and the results illustrate that there are positive correlations both in the absolute returns of actual data and predictive data.

INTRODUCTION

The objective of this work is to investigate the relationships between the crude oil market and the stock market and examine whether the shocks in crude oil price transmitted to Chinese stock market will receive considerable attention from investors. In the past decade, the crude oil demand of China is growing rapidly, and China has already become the second-largest oil importer in the world, after the United States. Fourteen years ago, China from an oil-exporting country became a net oil-importing country. From then on, the movement of crude oil prices had a strong influence on the economic behaviour of individuals and firms, and as a result, it affects the economic development directly. In another aspect, since July 2009, China has taken the place of Japan to be the world's second-largest stock market, and the stock market has played an important part in its economy. China has two stock markets: Shanghai Stock Exchange and Shenzhen Stock Exchange. The indices studied in the present paper are Shanghai Composite Index (SHCI) and Shenzhen Compositional Index (SZCI). These two most influential indices play an important role in Chinese stock markets. We also consider Shenzhen Petrochemical Index (SZPI) and the stock price of China's largest oil company: China Petroleum & Chemical Corporation (SINO-PEC). Daqing oil field and Shengli oil field are the first and the second largest oil fields in China respectively, the crude oil prices of Daqing and Shengli have a strong impact on Chinese energy market. The data for these crude oil prices and indices in the 7-year period is selected and analyzed by the statistical method and the neural network method.

Recently, some progress has been made in the study of fluctuations for the financial market and the energy market in China, for example see [1–7]. Artificial neural networks (ANNs) are one of the technologies that have made great progress in studying the stock markets [3, 8–11]. ANN have good self-learning ability, a strong ant jamming capability, and they have been widely used in financial fields such as stock prices, profits, exchange rate, and risk analysis and prediction. Although the historical data has a great influence on the investors' positions, we think that the impacts of different historical data on the stock price are not same. In the present paper, we suppose that the degree of impact of a data depends on its occurring date (or time), we give a high level effect of a data when it is very near to the current state. Furthermore, we also introduce the Brownian motion and Poisson jump in the model [3, 6, 11–15], in order to make the model have the effect of random movement and random jump while maintaining the original trend. In a financial market, jumps in financial assets play a crucial role in volatility forecasting. And jumps have a positive and mostly significant impact on future volatility. In this work, the artificial neural network model based on jump stochastic time effective function is applied to forecast the fluctuations of SHCI, SZCI, SZPI, Daqing, Shengli, and SINOPEC. We study the statistical behaviours and the linear regression for these indices, and the simulation plots and the comparisons of the observed data are given. We introduce mean absolute error (MAE), mean relative error (MRE), Theil's inequality coefficient (Theil's IC), bias proportion (BP), variance proportion (VP) and covariance proportion (CP) to evaluate the predictive results. Detrended fluctuation analysis (DFA) is developed to study both the stock markets and the crude oil markets [16–19]. DFA is one of the statistical analysis methods, which is applied to study the extent of long-range correlations in time series, it gives a statistical approach that reduces the effects of no stationary market trends and focuses on the intrinsic autocorrelation structure of market fluctuations over different time horizons. DFA provides a simple quantitative parameter, the scaling exponent α, to represent the correlation properties of time series. In the last part of Section 3, the empirical analysis shows the positive correlations in the absolute returns of the actual data and the predictive data by calculating the scaling exponent α.

In this paper, we introduce a new method: the jump stochastic time effective function in the neural network, to investigate the relationships between the crude oil market and the stock market. And the intelligent system, artificial neural networks with random theory is integrated in this work. The method is different from the methods used in previous papers [13, 14, 20], which also investigate the relationships between the crude oil market and the stock market. This paper also extends the method mentioned in [3] by introducing the random jump process, which can make the model have the effect of random jump while maintaining the original trend. And we do the different statistical analysis with the work in [3]. In the present paper, we improve the forecasting method in the neural network, each historical datum is given a weight (random with jump) depending on the time it occurs in the model, and we also use the probability density functions to classify the various variables from the training samples. The empirical research exhibits that the improved neural network model takes advantage over the traditional neural network models to some degree.

A BRIEF DESCRIPTION OF OIL MARKET AND STOCK MARKET IN CHINA

Chinese oil market is attracting more and more attentions from all over the world. China has been the world's second-largest oil consumer since 2003, and its oil demand reached 9% of the world's total demand in 2006. Figure 1 shows the monthly output and the monthly growth rate of the crude oil production in China from January 2003 to December 2009. The plot indicates that the crude oil output has

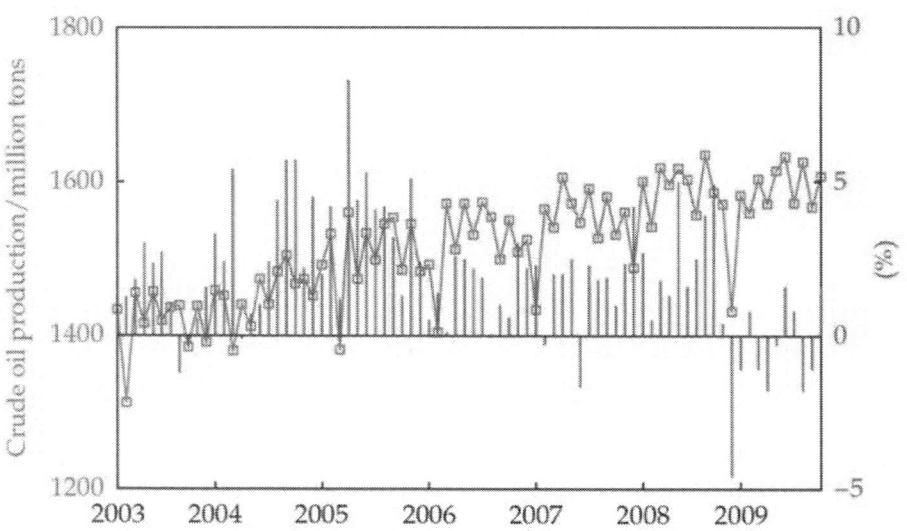

Figure 1: The output and the growth rate of crude oil in China.

almost reached the high limit, whereas the oil demand will grow by 4.5% in the coming three years. This displays that the stronger relationships between the international oil market and Chinese oil market become obvious.

In fact, China has become a net importer of crude oil since 1996; and the import dependence has exceeded 51%in 2008. Figures 2(a) and 2(b) present China's crude oil import and consumption monthly in the recent 7 years. The plots exhibit that the trends of the curves in Figures 2(a) and 2(b) are similar, which implies that the oil demand relies heavily on the international oil market. At the same time, the total values of China stock markets A shares reached 3.21 trillion US dollars on July 15; 2009, ranking as the world's second-largest stock market. The listed oil companies usually are the large cap companies, so the market capitalization value of these companies is not only a main part of the stock market value but also an important component of the stock market indices. Although some research work has been done in studying the relationship between the crude oil market and the stock market [4, 13, 14, 20–22], there has been relatively little empirical work done to analyze the relationships in China. In this paper, we select the data of SHCI, SZCI, SZPI, Daqing (Daqing crude oil price), Shengli (Shengli crude oil price), and the price of SINOPEC for each trading day in 7-year period from January 2, 2003 to December 31, 2009. And the corresponding statistical behaviors and comparisons of prices changes are studied in the following.

FORECASTING AND STATISTICAL ANALYSIS

In the real crude oil market, understanding the process by which oil prices evolve is fundamental to our knowledge of this market. Many empirical evidences, like the asymmetric and leptokurtic feature of return distributions and volatilities, strongly suggested an inappropriateness for the usage of Brownian motions in the Black-Scholes model. More precisely, it is often observed that the return distribution is skewed to zero and has a higher peak and fatter tails than those of the corresponding normal distribution. To explain those empirical phenomena, many

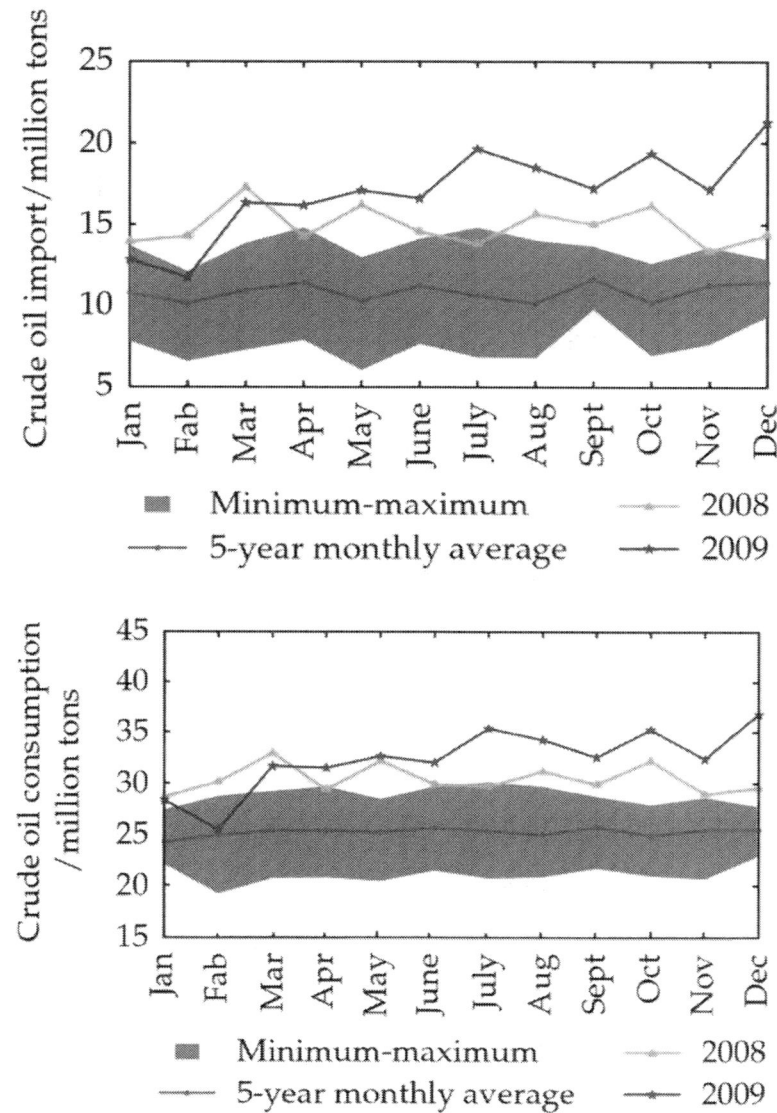

Figure 2 : (a) China's crude oil import, (b) China's crude oil consumption

researches propose innovative models such as normal jump diffusion models (see [12–15]), and continuous-time stochastic volatility models are becoming an increasingly popular way to describe moderate-and high-frequency financial data. These models introduce discontinuities, or jumps, into the volatility process, this can improve the empirical performance of these models. The distribution behaviour of jumps for oil prices often represents an important piece of the temporal crude oil price dynamics. We establish the presence of jumps in the data of the financial model, where the jumps that disrupt the entire term structure represent the most significant jump events. For example, in the present paper, these jump events may include the changing of international energy markets, the amount of oil production in China, and the crude oil reserve in China, Chinese oil consumption, Chinese energy policy, the wars, and the political events in the world, so on. These random events may be responsible for generating jumps in crude oil price dynamics. Since the fluctuation behaviours of the crude oil prices are also nonlinear, unstable, and random, we introduce the stochastic time effective function in the neural network. The function is supposed to follow a Brownian motion plus a compound Poisson process with a random jump distribution, in order to describe the above-mentioned empirical phenomena. We assume that the historical data of the crude oil market can reflect these random events, and affect the price volatility of the current oil market. For the model, the proposed stochastic time effective function may reflect the large fluctuations of the oil prices. Further, the function is a time-dependent random variable and also shows that the recent information has a stronger effect than the old information for the investors.

Jump Stochastic Time Effective Neural Network Model for Forecasting

There are various methods to forecast the volatilities of the time series, for example, the autoregressive conditional heteroscedasticity model has been applied by many financial analysts [23]. These financial time series models are based on the financial theories and require some strict assumptions on the distributions of the time series, so sometimes it is hard to reflect the market variables directly in the models. Usually stock prices can be seen as a random time sequence with noise, artificial neural networks, as large-scale parallel processing nonlinear systems that depend on their own intrinsic link data, providing methods and techniques that can approximate any nonlinear continuous function, without a priori assumptions about the nature of the generating process. The ANN model is a nonparametric method and can forecast future results by learning the pattern of market variables without any strict theoretical assumption [11]. Brooks demonstrated that it is applicable to forecast the volatilities of the financial time series by ANN [24].

First we introduce the three-layer BP neural network model in Figure 3, (for the details see [8–10]), and for any fixed neuron $n(n=1,2,...,N)$, the model has the following structure: let $\{xi(n):i=1,2,...,p\}$ denote the set of input of neurons, $\{yj(n):j=1,2,...,m\}$ denote the set of output of hidden layer neurons, Vi is weight that connects the node in i the input layer neurons to the node j in the hidden layer, Wj is weight that connects the node j in the hidden layer neurons to the node k in the output layer, and $\{ok(n):k=1,2,...,q\}$ denote the set of output of neurons. Then the output value for a unit is given by the following function

$$y_j(n) = f\left(\sum_{i=1}^{p} V_i x_i(n) - \theta_j\right), \quad o_k(n) = f\left(\sum_{i=1}^{p} W_j y_j(n) - \theta_k\right),$$

$$(3.1)$$

Where θj, θk are the neural thresholds, and $(x) = 1/(1+e-\alpha)$ is Sigmoid activation function. Let $Tk(n)$ be the actual value of data sets, then the error of the corresponding neuron k to the output is defined as $\varepsilon k=Tk-ok$.

Obviously, the real data follow normal distribution in general. However, the tail of the real distribution is fatter than the normal, which is called fat-tail phenomena. It is caused by drastic fluctuation of stock price. Moreover, we can find that the log return of stock price will fluctuate rapidly at intervals. In view of the above reality problem, the error of the output is defined as $\varepsilon = \varepsilon_k^2 / 2$ then the error of the sample $(n=1,2,...,N)$ is defined as

$$e(n,t) = \frac{1}{2} \phi(t) \sum_{k=1}^{q} (T_k(n) - o_k(n))^2,$$

$$(3.2)$$

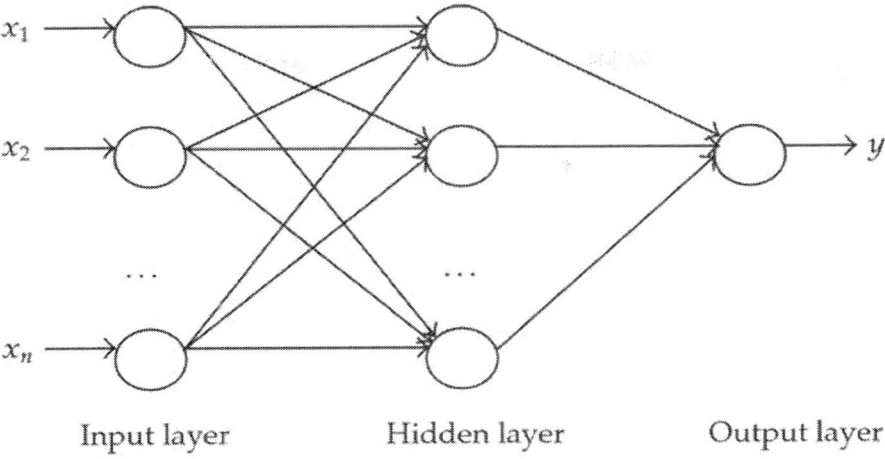

Input layer   Hidden layer   Output layer

Figure 3: The plot of three-layer neural network.

Where (t) is the jump stochastic time effective function. Now we defined (t) as follows

$$\phi(t_1 - t_n) = \frac{1}{\tau} \exp\left\{ -\int_{t_n}^{t_1} \mu(t)dt - \int_{t_n}^{t_1} \sigma(t)dB(t) + \sum_{l=1}^{N(t_1-t_n)} J_l \right\},$$

(3.3)

where $\tau(>0)$ is the time strength coefficient, $t1$ is the current time or the time of newest data in data set, and tn is an arbitrary time point in data set. Jl ($l=1, 2\ldots (t)$) Are independent and identically distributed jump processes and Jl obeying the normal distribution with mean μJ and variance σJ. N (t) $(t\geq0)$ is a Poisson process with intensity λ. (t) is the drift function (or the trend term), σ (t) is the volatility function, and (t) is the standard Brownian motion [5]. The stochastic time effective function implies that the recent information has a stronger effect for the investors than the old information. In detail, the nearer the events happened, the greater the investors and market are affected. Then the total error of all data training set in the set output layer with the jump stochastic time effective function is defined as1

$$E = \frac{1}{N}\sum_{n=1}^{N}e(n,t)$$

$$= \frac{1}{N}\sum_{n=1}^{N}\frac{1}{\tau}e^{-\int_{t_n}^{t_1}\mu(t)dt-\int_{t_n}^{t_1}\sigma(t)dB(t)+\sum_{i=1}^{N(t_1-t_n)}J_i}\sum_{k=1}^{q}\frac{1}{2}(T_k(n)-o_k(n))^2. \qquad (3.4)$$

Data is divided into two sections: the data from 2003 to 2007 is used for training and the rest is used for testing. For the stock indices, we input five kinds of stock prices: daily open price, daily closed price, daily highest price, daily lowest price, and daily trade volume, and one price of stock prices in the output layer: the closed price of the next trade day. And for the crude oil prices, we input five kinds of prices: the crude oil price of Brent, WTI, Dubai, Daqing, and Shengli, and the crude oil price of Daqing (or Shengli) of the next trade day is in the output layer. The number of neural nodes in input layer is 5, the number of neural nodes in the hidden layer is 13, and the number of neural nodes in output layer is 1. In this section, we take μJ and σJ to be the mean and the variance of reality historical data of SHCI, and let the intensity λ be 1/30. That is to say, jump will happen 10 times a year in average. Moreover, we suppose that the values of vector $((t), \sigma (t))$ are $(1, 1)$. The training algorithms procedures of the neural network are described as follows.

Step 1: Normalize the data as follows: $S(t)'=(S(t)-\min S(t))/(\max S(t)-\min S(t))$.

Step 2: At the beginning of data processing, connective weights Vi and Wj follow the uniform distribution on $(-1, 1)$, and let the neural threshold θk, θj be 0.

Step 3: Introducing the jump stochastic time effective function $\phi (t)$ in the error function $e (n, t)$. Choosing different volatility parameter. Giving the transfer function from input layer to hidden layer and the transfer function from hidden layer to output layer.

Step 4: Establishing an error-acceptable model and setting pre-set minimum error. If output error is below pre-set minimum error, go to Step 6, otherwise go to Step 5.

Step 5: Modify connective weights by calculating backward for the node in output layer:

$$\delta_o(n) = \frac{1}{\tau} \, e^{-\int_{t_n}^{t_1} \mu(t)dt - \int_{t_n}^{t_1} \sigma(t)dB(t) + \sum_{l=1}^{N(t_1-t_n)} I_l} o(n)[o(n) - T(n)][1 - o(n)]. \tag{3.5}$$

Calculate δ backward for the node in hidden layer:

$$\delta_h(n) = \frac{1}{\tau} \, e^{-\int_{t_n}^{t_1} \mu(t)dt - \int_{t_n}^{t_1} \sigma(t)dB(t) + \sum_{l=1}^{N(t_1-t_n)} I_l} o(n)[1 - o(n)] \sum_{h'} W_j \delta_{h'}(n), \tag{3.6}$$

where $o(n)$ is the output of the neuron n, $T(n)$ is the actual value of the neuron n in data sets, $o(n)[1-o(n)]$ is the derivative of the sigmoid activation function and h' is each of the node which connect with the node h and in the next hidden layer after node h. Modifying the weights from this layer to the previous layer:

$$W_j(n+1) = W_j(n) + \eta\delta_o(n)y(n) \quad \text{or} \quad V_j(n+1) = V_j(n) + \eta\delta_k(n)x(n), \tag{3.7}$$

Where η is learning step, which usually take constants between 0 and 1

Step 6: Output the predictive value.

Next, according to the computer simulations of the given neural network model, we do the comparisons between the predictive data of the model and the actual data of SHCI, SZCI, SZPI, Daqing, Shengli, and SINOPEC. And these comparison results are plotted in Figure 4.

In Figure 5, by using the linear regression method, we compare the predictive data of the neural network model with the actual data of SHCI, SZCI, SZPI, Daqing, Shengli, and SINOPEC. It is known that the linear regression attempts to model the relationship between two variables by fitting a linear equation to observed data. And it is usually used to fit a predictive model to an observed data set of two variables. Through the regression analysis, there are different linear equations in SHCI, SZCI, SZPI, Daqing, Shengli, and SINOPEC respectively, in Figure 5. We set the predictive data as x-axis and set the actual data as -axis, and the linear equation is $y=ax+b$. A valuable numerical measure of association between two variables is the correlation coefficient r. Table 1 shows the values of a, b, and r for the indices.

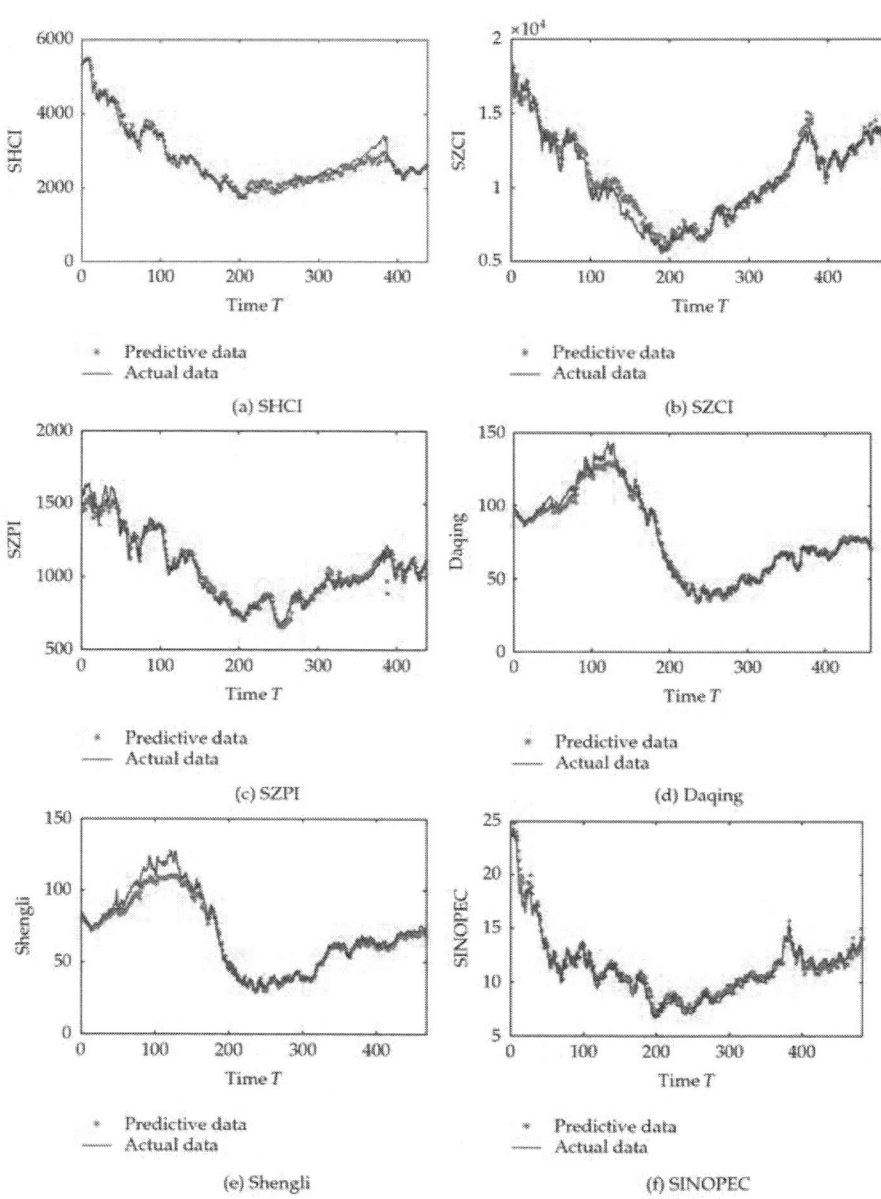

Figure 4: Comparisons of the predictive data and the actual data.

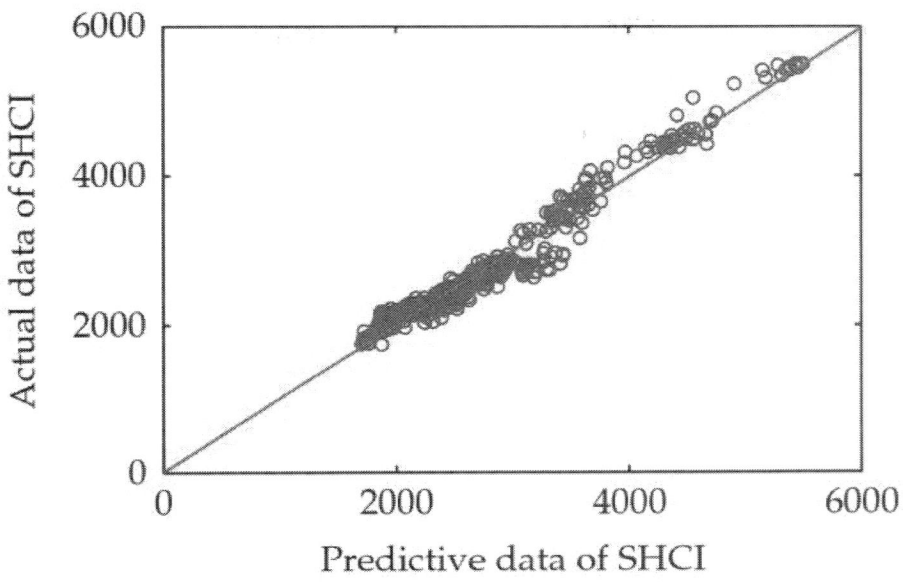

Figure 5: Regressions of the predictive data and the real data.

Table 1: Linear regression parameters

Parameter	SHCI	SZCI	SZPI	Daqing	Shengli	SINOPEC
a	0.9940	0.9715	0.9107	0.9217	0.8776	0.9934
b	8.5849	510.7603	91.7800	4.9774	5.6977	0.1957
r	0.9806	0.9824	0.9749	0.9942	0.9941	0.9792

Experiment Analysis

In Section 3.1, the financial price model is modeled by the neural network system. In order to evaluate the prediction of the model, we introduce some statistics in this section: mean absolute error (MAE), mean relative error (MRE), Theil inequality coefficient (Theil's IC), bias proportion (BP), variance proportion (VP) and covariance proportion (CP). We set $x_i, y_i, \bar{x}, \bar{y}, \sigma_x, \sigma_y$, and r as the predictive value, the actual value, the mean of the predictive value, the mean of the actual value, and the variance of the predictive value, the variance of the actual value and the correlation, respectively. These statistics are defined as follows:

$$\text{MAE} = \frac{1}{n}\sum_{i=1}^{n}|x_i - y_i|, \qquad \text{MRE} = \frac{1}{n}\sum_{i=1}^{n}\left|\frac{x_i - y_i}{y_i}\right|,$$

$$\text{Theil's IC} = \frac{\sqrt{(1/n)\sum_{i=1}^{n}(x_i - y_i)^2}}{\sqrt{(1/n)\sum_{i=1}^{n}x_i^2} + \sqrt{(1/n)\sum_{i=1}^{n}y_i^2}}, \tag{3.8}$$

Where the value of Theil IC is in [0, 1], and the smaller value means the better prediction of the model.

$$\text{BP} = \frac{(\bar{x} - \bar{y})^2}{\sum_{i=1}^{n}(x_i - y_i)^2/n}, \qquad \text{VP} = \frac{(\sigma_x - \sigma_y)^2}{\sum_{i=1}^{n}(x_i - y_i)^2/n},$$

$$\text{CP} = \frac{2(1 - r)\sigma_x\sigma_y}{\sum_{i=1}^{n}(x_i - y_i)^2/n} = 1 - \text{BP} - \text{VP}, \tag{3.9}$$

Where BP denotes the normalized difference between the mean of the predictive value and the mean of the actual value, and VP denotes the normalized difference between the variance of the predictive value and the variance of the actual value. Their values range from 0 to 1. The prediction of the model is effective when the value of CP is close to 1. Form the computer computation, Table 2 presents the values of the above statistics. Table 2 also gives a description of the deviating degrees between the predictive data and the actual data.

Table 2: Evaluation of the prediction.

	SHCI	SZCI	SZPI	Daqing	Shengli	SINOPEC
MAE	116.5679	433.4098	37.8812	2.8028	3.6815	0.4957
MRE	0.0431	0.0448	0.0343	0.0349	0.0466	0.0448
Theil's IC	0.0269	0.0260	0.0236	0.0348	0.0367	0.0273
BP	1.1760e-8	5.6445e-7	1.8592e-8	4.4732e-7	1.5301e-6	1.5276e-7
VP	0.0496	0.4514	0.15795	0.0047	0.0045	7.8972e-7
CP	0.9504	0.5486	0.84205	0.9942	0.9955	1.0000

In the next part, we will discuss the relationship between the crude oil price fluctuation of Daqing and the predictive values of the model. It is apparent in Figure 6(a) when the fluctuation is small; the predictive values are close to the actual values. In another aspect, when the fluctuation is large, the predictive values deviate from the actual values in some extent. We also can see in Figures 6(b) and 6(c) that the small fluctuation leads to the small relative errors and the small error bars

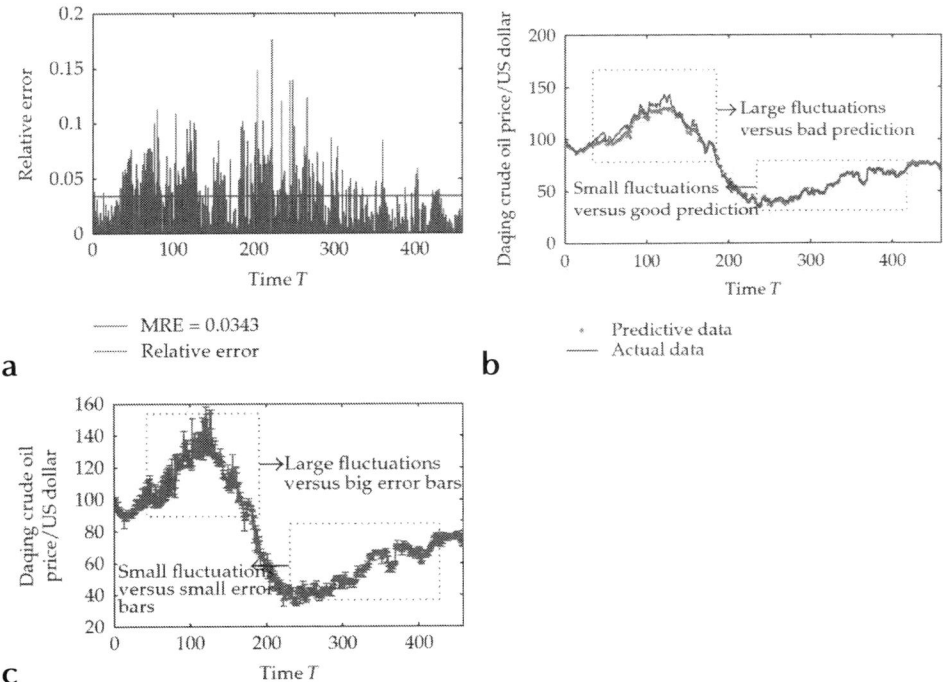

Figure 6: Comparisons of the fluctuation and the prediction of Daqing.

and the large fluctuation leads to the big relative errors and the big error bars. So there is a relationship between the fluctuation and the prediction. To investigate this relationship, we choose the predictive values and the actual values of Daqing as the research object. First, we measure the fluctuation in absolute returns, which is denoted by $|(t)|$. Then we divide the data into five groups by the absolute return intervals. The intervals are $[0, 0.01)$, $[0.01, 0.02)$, $[0.02, 0.03)$, $[0.03, 0.04)$, and [0.04, M], where M denotes the maximum of absolute returns. Table 3 shows the relationship between the actual fluctuation and the prediction by the absolute return intervals.

Table 3: The relationship between the fluctuation and the prediction by the absolute return intervals.

$| R(t) |$ intervals	MAE	MRE
[0,0.01)	0.0312	17.3123
[0.01,0.02)	0.0363	16.4007
[0.02,0.03)	0.0401	16.3452
[0.03,0.04)	0.0387	14.8879
[0.04,M]	0.0376	26.2455
[0,M]	0.0382	18.2984

Return Analysis

In this section, we discuss the statistical properties of SHCI, SHZI, SZPI, Daqing, Shengli, and SINOPEC in the 7-year period from January 2003 to December 2009. Figure 7 presents the figures of the returns time sequence for these indices. We denote the daily price at time t by (t)

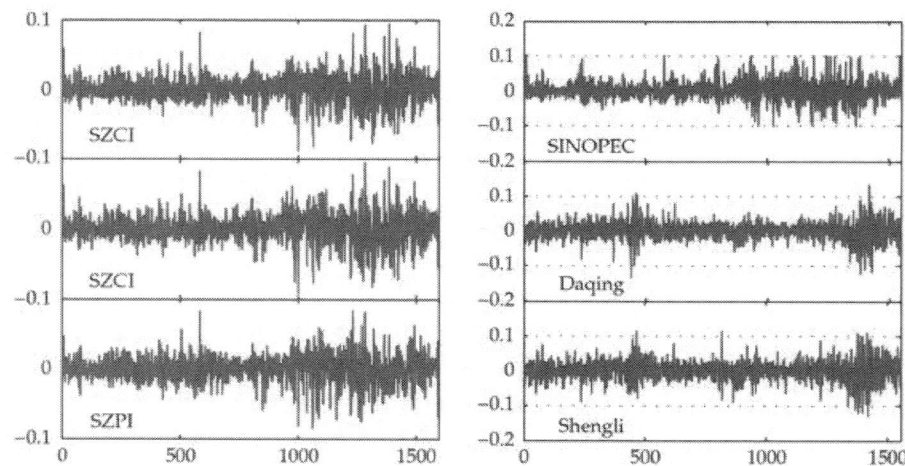

Figure 7: Returns of the indices in the 7-year period from January 2003 to December 2009.

$(t=0, 1, 2...)$, then the return of the stock price (or index) is given by

$$R(t) = \frac{S(t+1) - S(t)}{S(t)} = \frac{S(t+1)}{S(t)} - 1.$$

(3.10)

Table 4 presents the statistical analysis of the returns for the actual data. Note that the daily price fluctuation is limited in China, that is, the changing limits of the daily returns for stock prices and stock indices are between 10% and −10%, whereas the returns of the crude oil price can change in a larger value range. Table 5 presents the statistical analysis of the returns for the predictive data. In these two tables, they show the values of mean, variance, kurtosis and skewness of the returns, and we also can compare these values between the actual data and the predictive data.

Table 4: Returns statistics of the real data.

	SHCI	SZCI	SZPI	Daqing	Shengli	SINOPEC
Mean	7.6353e−4	1.2217e−3	8.7236e−4	6.8830e−4	8.4244e−4	1.5475e−3
Variance	3.3633e−4	3.9532e−4	3.6080e−4	6.3243e−4	7.9565e−4	7.4543e−4
Skewness	−0.0779	−0.1274	−0.3832	−0.1276	−0.0643	0.3245
Kurtosis	3.1675	2.4920	2.2832	3.1553	1.8875	2.4495
Minimum	−0.0884	−0.0932	−0.0844	−0.1352	−0.1263	−0.1030
Maximum	0.0954	0.0963	0.0843	0.1323	0.1146	0.1015

Table 5: Returns statistics of the predictive data

	SHCI	SZCI	SZPI	Daqing	Shengli	SINOPEC
Mean	−9.9940e−4	−4.3613e−5	2.2083e−4	−7.4974e−5	−2.0295e−4	−4.9588e−4
Variance	8.3044e−4	1.1969e−3	9.5615e−4	1.0991e−3	1.1038e−3	1.2144e−3
Skewness	0.6257	0.0838	1.0189	0.2120	0.1417	0.5700
Kurtosis	2.6270	2.3022	1.9964	3.1030	2.0811	1.7350
Minimum	−0.08211	−0.1327	−0.2057	−0.1227	−0.1101	−0.1073
Maximum	0.1525	0.1340	0.2815	0.1589	0.1250	0.1597

Detrended Fluctuation Analysis

Detrended fluctuation analysis (DFA) is a scaling analysis method providing the scaling exponent α to represent the correlation properties [7,

16–18]. There are two advantages in DFA method. One is that it permits the detection of long-range correlations embedded in seemingly no stationary time series. The other is that it avoids the spurious detection of apparent long-range correlations that are artifact of nonstationarity. Briefly, for a given stochastic time series (i), $i=1,2,...,N$, with the sampling period Δt, the DFA method can be implemented as follows.

Step 1: Compute the mean $\bar{S} = (1/N)\sum_{i=1}^{N} S(i)$ and obtain an integrated time series $y(j) = (1/N)\sum_{i=1}^{j}(S(i) - \bar{S})$.. Then divide the integrated time series into boxes of equal size, n.

Step 2: In each box, fit the integrated time series by using a polynomial function, $yfit(i)$. For order-l DFA, $l$order polynomial function should be applied for the fitting and in this paper, $l=2$. Then calculate the detrended fluctuation function as follows:

$$Y(i) = y(i) - y_{fit}(i). \qquad (3.11)$$

Step 3: For a given box size n, calculate the root mean square fluctuation:

$$F(n) = \left(\frac{1}{N} \sum_{i=1}^{N} [Y(i)]^2 \right)^{1/2} \qquad (3.12)$$

A power-law relation between (n) and the box size n indicates the presence of scaling: $(n) \sim n\alpha$. The parameter α, called the scaling exponent or correlation exponent, represents the correlation properties of the time series: if $\alpha=0.5$, there is no correlation and the time series is uncorrelated; if $\alpha<0.5$, the signal is anticorrelated; if $\alpha>0.5$, there are positive correlations in the time series.

In this paper, we use DFA to analyze the absolute returns of the actual data and the predictive data, see Figure8. αA and αP denote the scaling exponents of the absolute returns for the actual data and the predictive data respectively. Table 6 shows that αA and αP are all larger than 0.5, which means that there are positive correlations in the absolute returns of the actual data and the predictive data.

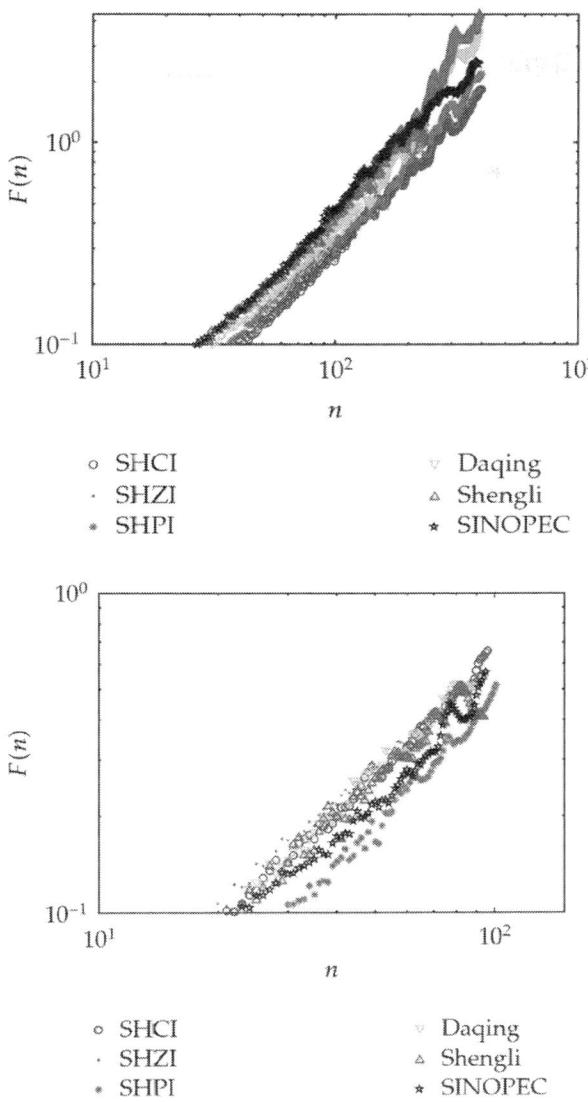

Figure 8: Detrended fluctuation analysis for the absolute returns of the actual data and the predictive data. (a) The plot of the absolute returns for the actual data from January 2003 to December 2009. (b) The plot of the absolute returns of the predictive data from January 2008 to December 2009.

Table 6: Scaling exponent of the absolute returns.

Scaling exponent	SHCI	SZCI	SZPI	Daqing	Shengli	SINOPEC
αA	3.5852	3.6523	3.7007	3.8456	3.9760	3.6342
αP	5.6438	5.7443	5.6002	5.9843	5.7712	5.8732

CONCLUSION

In this paper, we introduce the jump stochastic time effective neural network model to forecast the fluctuations of SHCI, SZCI, SZPI, Daqing, Shengli, and SINOPEC. The corresponding statistical behaviors of these indices are investigated; and several kinds of comparisons between the actual data and the predictive data are given. Further, the absolute returns of the actual data and the predictive data are studied by the statistical method and the detrended fluctuation analysis.

ACKNOWLEDGMENTS

The authors were supported in part by National Natural Science Foundation of China Grant nos. 70771006 and 10971010, and BJTU Foundation grant no. S11M00010.

REFERENCES

1. R. G. Cong, Y. M. Wei, J. L. Jiao, and Y. Fan, "Relationships between oil price shocks and stock market: an empirical analysis from China," Energy Policy, vol. 36, no. 9, pp. 3544–3553, 2008.

2. M. F. Ji and J. Wang, "Data analysis and statistical properties of Shenzhen and Shanghai land indices,"WSEAS Transactions on Business and Economics, vol. 4, pp. 33–39, 2007.

3. Z. Liao and J. Wang, "Forecasting model of global stock index by stochastic time effective neural network," Expert Systems with Applications, vol. 37, no. 1, pp. 834–841, 2010.

4. T. C. Mills, The Econometric Modelling of Financial Time Series, Cambridge University Press, Cambridge, UK, Second edition, 1999.

5. J. Wang, Stochastic Process and Its Application in Finance, Tsinghua University Press and Beijing Jiaotong University Press, Beijing, China, 2007.

6. J. Wang, Q. Wang, and J. Shao, "Fluctuations of stock price model by statistical physics systems,"Mathematical and Computer Modelling, vol. 51, no. 5-6, pp. 431–440, 2010.

7. T. Wang, J. Wang, and B. Fan, "Statistical analysis by statistical physics model for the stock markets,"International Journal of Modern Physics C, vol. 20, no. 10, pp. 1547–1562, 2009.

8. E. M. Azoff, Neural Network Time Series Forecasting of Financial Market, Wiley, New York, NY, USA, 1994.

9. M. Demuth and M. Beale, Neural Network Toolbox: For Use with MATLAB, The Math Works, Inc., Natick, Mass, USA, 5th edition, 1998.

10. V. S. Desai and R. Bharati, "The efficacy of neural networks in predicting returns on stock and bond indices," Decision Sciences, vol. 29, no. 2, pp. 405–423, 1998.

11. T. Hyup Roh, "Forecasting the volatility of stock price index," Expert Systems with Applications, vol. 33, no. 4, pp. 916–922, 2007.

12. O. B. Nielsen and N. Shephard, "Power and bipower variation with stochastic volatility and jumps,"Journal of Financial Econometrics, vol. 2004, no. 2, pp. 1–48, 2004.

13. U. Oberndorfer, "Energy prices, volatility, and the stock market: evidence from the Eurozone," Energy Policy, vol. 37, no. 12, pp. 5787–5795, 2009.

14. E. Papapetrou, "Oil price shocks, stock market, economic activity and employment in Greece," Energy Economics, vol. 23, no. 5, pp. 511–532, 2001.

15. D. Pirino, "Jump detection and long range dependence," Physica A, vol. 388, no. 7, pp. 1150–1156, 2009.

16. J. Alvarez-Ramirez, J. Alvarez, and E. Rodriguez, "Short-term predictability of crude oil markets: a detrended fluctuation analysis approach," Energy Economics, vol. 30, no. 5, pp. 2645–2656, 2008.

17. O. F. Ayadi, J. Williams, and L. M. Hyman, "Fractional dynamic behavior in Forcados Oil Price Series: an application of detrended fluctuation analysis," Energy for Sustainable Development, vol. 13, no. 1, pp. 11–17, 2009.

18. K. Hu, P. C. Ivanov, Z. Chen, P. Carpena, and H. E. Stanley, "Effect of trends on detrended fluctuation analysis," Physical Review E, vol. 64, no. 1, pp. 0111141–01111419, 2001.

19. Y. Wang, L. Liu, and R. Gu, "Analysis of efficiency for Shenzhen stock market based on multifractal detrended fluctuation analysis," International Review of Financial Analysis, vol. 18, no. 5, pp. 271–276, 2009.

20. S. Saif Ghouri, "Assessment of the relationship between oil prices and US oil stocks," Energy Policy, vol. 34, no. 17, pp. 3327–3333, 2006.

21. R. Gaylord and P. Wellin, Computer Simulations with Mathematica: Explorations in the Physical, Biological and Social Science, Springer, New York, NY, USA, 1995.

22. K. Ilinski, Physics Of Finance: Gauge Modeling in Non-Equilibrium Pricing, John Wiley, New York, NY, USA, 2001.

23. R. F. Engle, "Autoregressive conditional heteroscedasticity with estimates of the variance of United Kingdom inflation," Econometrica, vol. 50, no. 4, pp. 987–1007, 1982.

24. C. Brooks, "Predicting stock index volatility: can market volume help?" Journal of Forecasting, vol. 17, pp. 59–80, 1998.

CITATION

Jun Wang, Huopo Pan, and Fajiang Liu, "Forecasting Crude Oil Price and Stock Price by Jump Stochastic Time Effective Neural Network Model," doi:10.1155/2012/646475

Model by Lattice Fractal Sierpinski Carpet Percolation

Xu Wang and Jun WangDepartment
Department of Mathematics, Key
Laboratory of Communication and
Information System, Beijing Jiaotong
University, Beijing 100044, China

9

ABSTRACT

The lattice fractal Sierpinski carpet and the percolation theory are applied to develop a new random stock price for the financial market. Percolation theory is usually used to describe the behavior of connected clusters in a random graph, and Sierpinski carpet is an infinitely ramified fractal. In this paper, we consider percolation on the Sierpinski carpet lattice, and the corresponding financial price model is given and investigated. Then, we analyze the statistical behaviors of the Hong Kong Hang Seng Index and the simulative data derived from the financial model by comparison.

INTRODUCTION

Financial fluctuation system is one of complex systems, and the statistical behavior of fluctuation of stock price changes has long been a focus of financial research. With the flourishing research of complex systems, it becomes more and more attractive to find universal rules and principles of these systems and further to answer the origination of financial complex system. Recent research is no longer restricted to the traditional areas but concentrated on the more comprehensive domains, leading to the birth of many burgeoning disciplines through the interaction and amalgamation of mathematics and other fields such as finance, biology, and sociology. For example, the theory of stochastic interacting particle systems (see [1–6]) recently has been applied to study the behaviors of market fluctuations, see [7–15]. And the study of financial market prices has been found to exhibit some universal properties similar to those observed in interacting particle systems with a large number of interacting units.

Percolation theory, as a model (in interacting particle systems) for a disordered medium, has brought new understanding and techniques to a broad range of topics in nature and society. First we consider the bond percolation on Zd, that is, for $x,y \in Zd$, the distance $\delta(x,y)$ from x to y is defined by $\delta(x,y)=\sum_{i=1}^{d}|x_1 - y_i|$, where $x=(x1,...,xd)$ and $y=(y1-,yd)$. By adding edges (or bonds) between all pairs x, y of points of Zd with $(x,)=1$, we establish the d-dimensional lattice $\mathbb{L}d=(Zd,\mathbb{E}d)$, and we write $\mathbb{E}d$ for the set of the edges. Suppose that each bond of lattice $\mathbb{L}d$ is either open (occupied) with probability p or closed (empty) with probability $1-p$, then connected components of this graph are called open clusters. Let (x) denote the open cluster containing the vertex x, and $(p)=P(|C(0)|=\infty)$ be the probability that the origin belongs to an infinite open cluster. When the intensity p increases from zero to one, at some sharp percolation threshold (or critical point) pc, for the first time, one infinite cluster appears; for all $p>pc$ we have exactly one infinite cluster, for all $p<pc$ we have no infinite cluster, and at critical value $p=pc$ the incipient infinite clusters are supposed to be fractal.

A lattice fractal is a graph which corresponds to a fractal, all of them have a self-similarity, but most of them have no translation invariance, see [1, 16–19]. The Sierpinski gasket and the Sierpinski carpet are well-known examples of fractals. The former is a finitely ramified fractal (i.e., it can be disconnected by removing a finite number of points) and the latter is an infinitely ramified fractal. Fractals also have close relations to financial markets [17], electrical conductivity, superconductivity, and mechanical properties of percolating systems, and so forth. In [1], it shows that the Ising model on the lattice Sierpinski carpet does exhibit the phase transition in any dimension, but the Ising model on the lattice Sierpinski gasket has no phase transition in any dimension (because of the character of the finitely ramified fractal). Similar results of phase transitions can be obtained for percolation on the lattice Sierpinski carpet and on the lattice Sierpinski gasket, see [18].

In the present paper, a new method is introduced to model and describe the fluctuations of market prices, namely, we use the lattice fractal Sierpinski carpet percolation to establish a new random market price in a financial market. In this financial model, the local interaction or influence among traders in one stock market is constructed, and a cluster of percolation is used to define the cluster of traders sharing the same opinion about the market. For the comparison, we also consider the most important index of Hong Kong financial market, the Hong Kong Hang Seng Index. We analyze the statistical properties of Hong Kong Hang Seng Index and the simulative data derived from the price model by comparison, which including the sharp peak and the fat-tail distribution for the price changes, the distribution of returns decays with power law in the tails, the price fluctuations are not invariant against time reversal (i.e., they show a forward-backward asymmetry), and so forth. Moreover, the behaviors of long memory and long-range correlation in volatility series of market returns are exhibited.

DESCRIPTION OF PRICE MODEL ON LATTICE SIERPINSKI CARPET PERCOLATION

First we give a brief description of percolation on the lattice Sierpinski carpet (d) (for $d=2$), which is defined as follows: consider Z2 as a graph in the usual sense and set

$$\widetilde{\mathbb{S}}_{n+1}^{(2)} = \bigcup_{\substack{i_1,i_2 \in \{0,1,2\} \\ (i_1,i_2) \neq (1,1)}} \left\{ \left(i_1 3^{n+1}, i_2 3^{n+1} \right) + \widetilde{\mathbb{S}}_n^{(2)} \right\},$$

(2.1)

where

$$u + \widetilde{\mathbb{S}}_n^{(2)} = \{ u + v : v \in \widetilde{\mathbb{S}}_n^{(2)} \}$$

To make the graph more symmetric, let $\mathbb{S}_n^{(2)}$ be the union of $\widetilde{\mathbb{S}}_n^{(2)}$ and its reflections in every coordinate hyperplane. Then we define the lattice Sierpinski carpet as

$$\mathbb{S}^{(2)} = \bigcup_{n=0}^{\infty} \mathbb{S}_n^{(2)}.$$

(2.2)

Similarly to Section 1, we define the corresponding edges set of (2) as ((2)). Next we consider random graph (bond percolation) on the lattice $((2))=(\mathbb{S}(2),\mathbb{E}(\mathbb{S}(2)))$, see Figure 1. Let p (the intensity value) satisfies $0 \leq p \leq 1$, each edge of ((2)) is declared to be open with probability p and closed with probability $1-p$ independently. We denote the product probability by Pp (or P), and define $(p)=P(|C(0)|=\infty)$, where $C(0)$ is the open cluster containing the origin on $\mathbb{L}(\mathbb{S}(2))$, and $|C(0)|$ is the number of vertices in $C(0)$. Let $pc(\mathbb{S}(2))=\inf\{p:\theta(p)>0\}$, then percolation on the Sierpinski carpet $\mathbb{S}(2)$ exhibits the existence of a phase transition, that is, $\theta(p)>0$ for $p>pc(\mathbb{S}(2))$, for details see [1, 18].]

Next we consider a price model of auctions for a stock in a stock market. Assume that each trader can trade the stock several times at each day $t \in \{1,2,...,T\}$, but at most one unit number of the stock at each time. Let(t) denote the daily closing price of tth trading day. And let Λn be a subset of (2),

where

$$\Lambda_n = \left\{ (x_1, x_2) \in \mathbb{S}^{(2)} : -3^n \leq x_1 \leq 3^n, -3^n \leq x_2 \leq 3^n \right\}$$

(2.3)

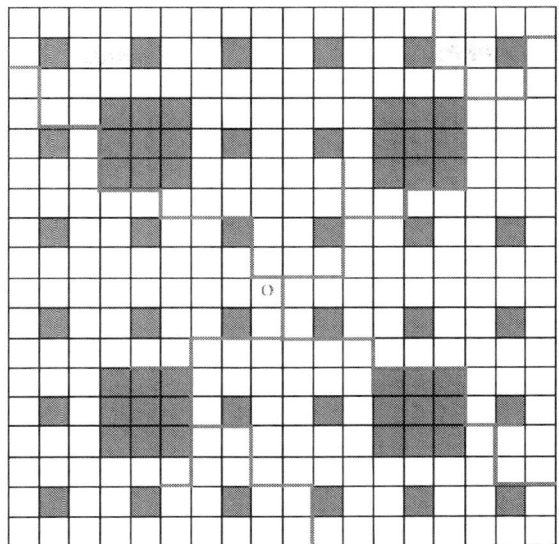

Figure 1: Lattice percolation on lattice Sierpinski carpet.

and $C_t(0)$ be a random open cluster on Λ_n. Suppose that this stock consists of $|\Lambda_n|$ (n is large enough) investors, who are located in Λ_n lattice. And (0) is a random set of the selected traders who receive the information. At the beginning of trading in each day, suppose that the investors receive some news. We define a random variable ζ_t for these investors, suppose that these investors taking buying positions ($\zeta_t=1$) selling positions ($\zeta_t=-1$), or neutral positions ($\zeta_t=0$) with probability $q1, q-1$ or $1-(q1+q-1)(q1, q2>0, q1+q2\leq 1)$, respectively. Then these investors send bullish, bearish or neutral signal to the market. According to bond percolation on (2), investors can affect each other or the news can be spread, which is assumed as the main factor of price fluctuations. For a fixed $t\in\{1,2,\ldots,T\}$, let

$$B_t = \frac{\zeta_t|C_t(0)|}{|\Lambda_n|}.$$

(2.4)

From the above definitions and mathematical finance theory [20–24], we define the stock price at t th trading day as

$$S(t) = e^{\alpha(t)B_t} S(t-1),$$

$$(2.5)$$

where $S(0)$ is the initial stock price at time 0, and $\alpha(t)$ (>0) represents the depth function of the market at trading day t. Then we have

$$S(t) = S(0) \exp\left\{ \alpha(t) \sum_{k=1}^{t} B_k \right\}, \quad t \in \{1, 2, \ldots, T\}.$$

$$(2.6)$$

The formula of the single-period stock logarithmic returns from t to $t+1$ is given as follows:

$$r(t) = \ln S(t+1) - \ln S(t), \quad t \in \{1, 2, \ldots, T\}.$$

$$(2.7)$$

EXPERIMENT ANALYSIS OF MARKET RETURN DISTRIBUTION

In order to make empirical research on the financial price model and an actual stock market by comparison, we select the daily closing prices of Hang Seng Index in the 20-year period from September 3, 1990 to September 3, 2010, the total number of observed data is about 4942. Recent research shows that returns on financial markets are not Gaussian, but exhibit excess kurtosis and fatter tails than the normal distribution, which is usually called the "fat-tail" phenomenon, see [21, 25–30]. The general explanation for this phenomenon is thought to be the "herd effect" of investors in the market. The time series of returns by simulating the price model which is developed on the Sierpinski carpet percolation is plotted in Figure 2(a). The returns distributions of Hang Seng Index and the financial model are plotted in Figure 2(b), the part (where the probability is above the 75th or below 25th percentiles of the samples) deviates from the dash line. This implies that the probability distributions of returns deviate from the corresponding normal distributions at the tail parts.

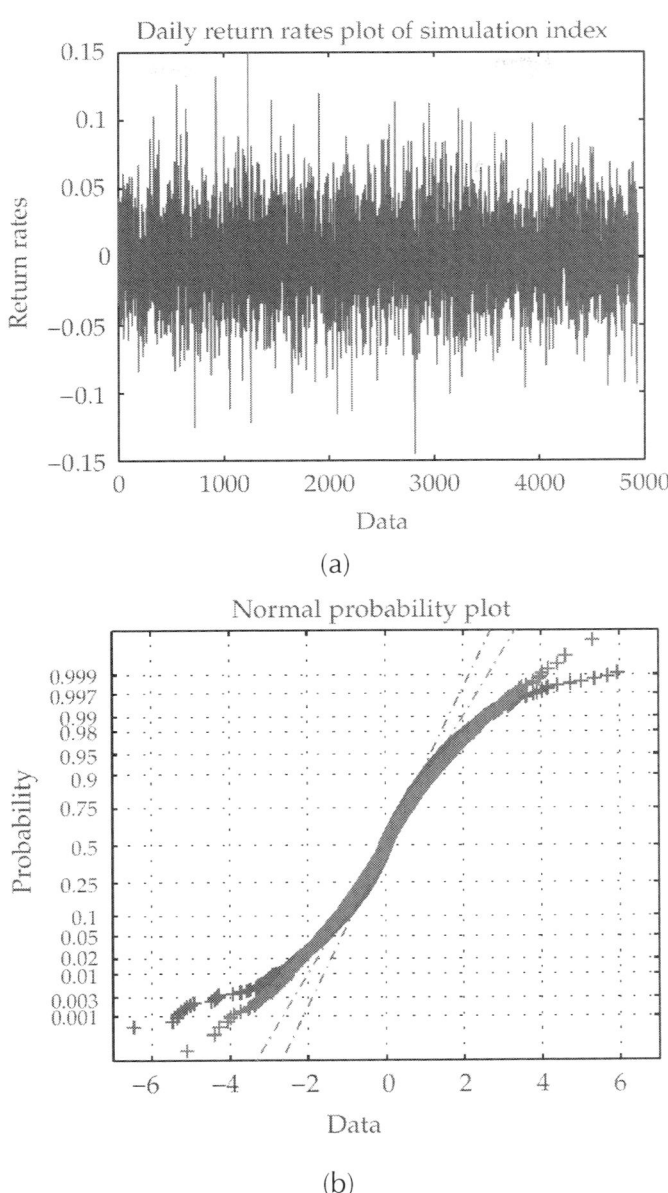

Figure 2: (a) The returns time series of simulation data for the price model with the intensity $p=0.49$. (b) The comparison of returns distributions for 20-year period Hang Seng index and the simulative data with $p=0.49$.

For further analyzing the character of returns distributions for the simulative data and Hang Seng Index, we make the single-sample Kolmogorov-Smirnov test by the statistical method, the basic statistics of the corresponding returns is displayed in Table 1. The value of two-tail test P is 0.000, thus the hypothesis is denied that the distribution of returns follows the Gaussian distribution.

Table 1: The Kolmogorov-Smirnov test

	The financial model	Hang Seng Index
Capability	4942	4942
The H value	1	1
The P value of double tail	0.0000	0.0000
K-S statistics to measure	0.4633	0.4736
The CV value	0.0193	0.0193

In this part, we study the properties of skewness and kurtosis on the returns for the simulative data and Hang Seng Index. Next we give the definitions of skewness and kurtosis as follows:

$$\text{Skewness} = \sum_{i=1}^{n} \frac{(r_i - u_r)^3}{(n-1)\delta^3},$$

$$\text{Kurtosis} = \sum_{i=1}^{n} \frac{(r_i - u_r)^4}{(n-1)\delta^4},$$

(3.1)

where r_i denotes the return of ith trading day, u_r is the mean of r, n is the total number of trading dates, and δ is the corresponding standard variance. Kurtosis shows the centrality of data, and the skewness shows the symmetry of the data; it is a measure of the "peakedness" of the probability distribution of a real-valued random variable, and the infrequent extreme deviations lead higher kurtosis. Skewness is important because kurtosis is not independent of skewness, and the latter may "induce" the former. It is known that the skewness of standard normal distribution is 0 and the kurtosis is 3. Next we investigate the statistical behaviors of the returns for different intensity values p, where the value p changes from 0.39 to 0.55 with the interval length 0.01.

Table 2 gives a description of the statistics for 17 group data of the price model. This shows that the distribution of the returns deviates from the Gaussian distribution with the intensity values p increasing, and the kurtosis distribution of the returns has a sharper peak, longer and fatter tails for larger p. From the definitions in Section 2, p is the intensity for the Sierpinski carpet lattice percolation and represents the strength of information spread in the price model. The wider the information spread, the larger the value of p is. In the following, we hope to exhibit that the numerical characteristics of simulation results for some intensity p are very close to those of the real data. We analyze the probability distributions of the logarithmic returns and the cumulative distributions of the normalized returns for these data in Figure 3, where the intensity values of the model are $p=0.485$, $p=0.49$, and $p=0.495$, respectively.

Table 2: The analysis of the price model for different intensity values p

p	Kurtosis	Skewness	Mean	Variance	Min	Max
0.55	5.144784	-0.09809	$-9.88E-06$	$2.19E-07$	-0.00242	0.002424
0.54	5.131352	-0.02482	$-7.67E-06$	$2.80E-07$	-0.0029	0.003056
0.53	4.893179	0.02581	$2.39E-06$	$3.35E-07$	-0.00316	0.003425
0.52	5.145840	-0.12553	$1.41E-06$	$4.17E-07$	-0.00411	0.003056
0.51	4.923338	-0.03637	$1.16E-05$	$5.47E-07$	-0.00379	0.003741
0.50	4.417513	0.061596	$6.37E-06$	$6.88E-07$	-0.00374	0.004057
0.49	5.114434	-0.10168	$-6.15E-06$	$8.47E-07$	-0.00559	0.004321
0.48	4.697023	0.030538	$-1.25E-05$	$1.11E-06$	-0.00495	0.005954
0.47	4.484873	-0.07156	$-1.90E-05$	$1.47E-06$	-0.00669	0.005427
0.46	5.045552	0.13255	$1.03E-05$	$1.92E-06$	-0.00801	0.00743
0.45	5.066152	-0.07007	$-3.35E-05$	$2.46E-06$	-0.01112	0.008536
0.44	4.491076	0.076139	$2.90E-05$	$3.11E-06$	-0.00843	0.008325
0.43	4.905725	-0.14384	$5.43E-05$	$4.61E-06$	-0.01565	0.008115
0.42	5.029006	-0.03238	$-1.70E-06$	$6.18E-06$	-0.01655	0.013437
0.41	4.905482	-0.03227	$-2.31E-05$	$8.84E-06$	-0.01813	0.013806
0.40	4.539316	-0.05757	$1.30E-05$	$1.23E-05$	-0.01739	0.01702
0.39	4.431923	0.042044	$-6.44E-06$	$1.92E-05$	-0.02176	0.019338

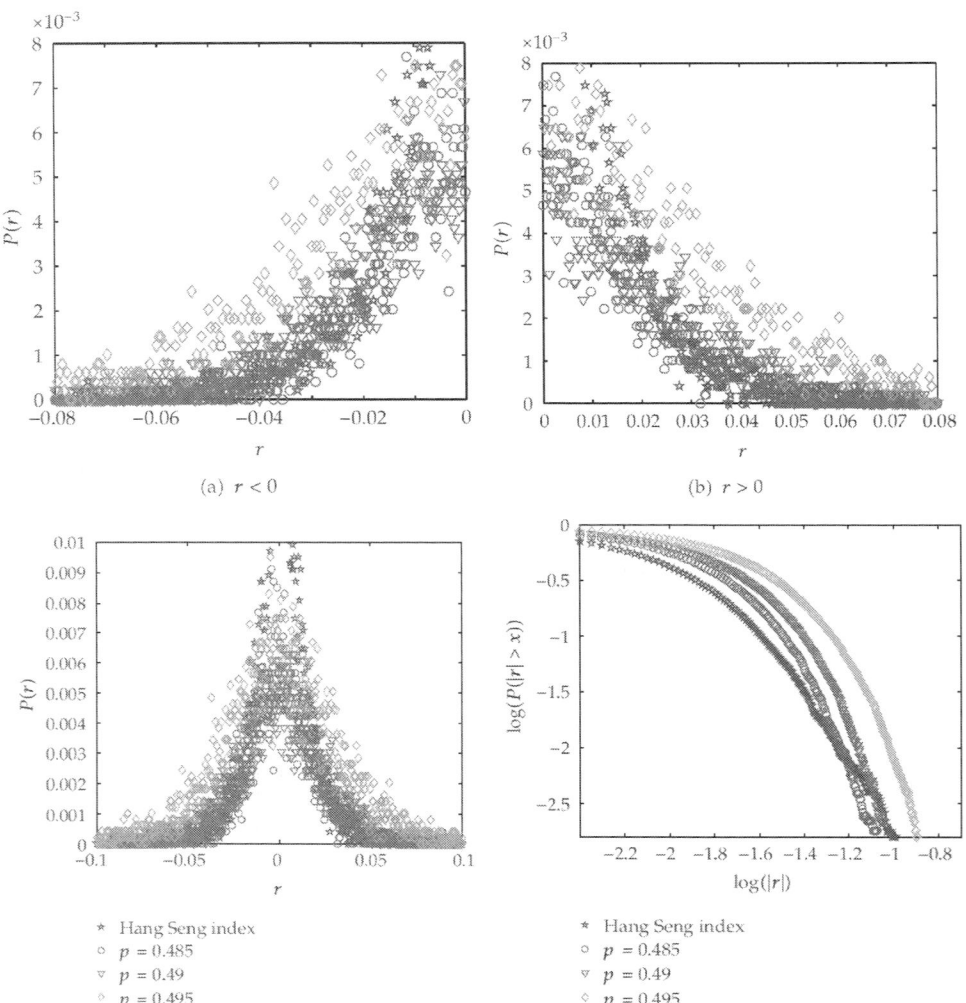

Figure 3: The plots (a), (b), and (c) are the probability distributions of the logarithmic returns, and the plot (d) is the cumulative distributions of the normalized price returns. The data is selected from Hang Seng Index and from the simulation data with the different values p, $p=0.485$, $p=0.49$ and $p=0.495$.

LONG MEMORY TEST OF THE MODEL AND HANG SENG INDEX

We analyze the long memory of the returns by using Lo's modified res-caled range statistic [31]. The long memory is measured by the Hurst exponent H, calculated by Lo's modified rescaled range statistic. For $0.5 < H < 1$, the series exhibits the long-term persistence, with the maintenance of tendency; for $0 < H < 0.5$, the series is the antipersistent, presenting reversion to the mean; and for $H = 0.5$, the series corresponds to a random walk. We consider a sample of series $X1,2,\ldots,Xn$ and let $\overline{X}_n$ denote the sample mean. Then the modified rescaled range statistic, denoted by Qn, is defined by

$$Q_n = \frac{1}{\hat{\sigma}_n(q)}\left[\max_{1\le k\le n}\sum_{j=1}^{k}\left(X_j - \overline{X}_n\right) - \min_{1\le k\le n}\sum_{j=1}^{k}\left(X_j - \overline{X}_n\right)\right],$$

(4.1)

where

$$\hat{\sigma}_n^2(q) = \frac{1}{n}\sum_{j=1}^{n}\left(X_j - \overline{X}_n\right)^2 + \frac{2}{n}\sum_{j=1}^{q}w_j(q)\left[\sum_{i=j+1}^{n}\left(X_i - \overline{X}_n\right)\left(X_{i-j} - \overline{X}_n\right)\right]$$

$$= \hat{\sigma}_X^2 + 2\sum_{j=1}^{q}w_j(q)\hat{\gamma}_j,$$

(4.2)

$$w_j(q) = 1 - \frac{j}{q+1}(q < n)$$

$\hat{\sigma}_X^2$ and $\hat{\gamma}_i$ denote the sample variance and the autocovariance estimators of X.

In order to make the statistical inference for the above-modified res-caled range statistics, Lo derived that $(q)=n1/2Qn$ converges weakly to a random variable V, where V is the range of a Brownian bridge on the unit interval. Then the corresponding distribution function of V is given by

$$F(v) = 1 + 2\sum_{k=1}^{\infty}\left(1 - 4k^2v^2\right)e^{-2(kv)^2}.$$

(4.3)

Form this function (v), we can get test threshold for any level of signifi-cance (by examining significant of (q)), this reflects the long memory behavior for the time series. It is important to select the window wide q; we take the experience value

$$q = \left(\frac{3T}{2}\right)^{1/3} \cdot \left(\frac{2\hat{\rho}_1}{1 - \hat{\rho}_1^2}\right)^{2/3},$$

$$(4.4)$$

Where $\hat{\rho}_1$ is the estimation of first-order autocorrelation coefficient of the time series. Then the Hurst exponent H is defined as the limit of the ratio $\log Q_n/\log n$. At the same time, it shows a linear growth trend between modified R/S statistic and sample size n, by using regressionln

$$\ln Q_n = \ln c + H \ln n.$$

$$(4.5)$$

With some optimal q value, the statistics of returns by the modified R/S statistic is given in Table 3 and Figure 4. Figure 4 also shows the fluctuations of exponent H of returns for the price model and Hang Seng Index.

Table 3: Statistics of returns for V and H

Return series	V statistic results				Hurst index results	
	n	First-order autocorrelation	q	V	Intercept c	H
The financial model	4942	0.018461152	2.1624	1.3901	0.129960913	0.5175
Hang Seng Index	4942	0.009450232	1.3835	1.2361	−0.357078997	0.5893

LONG-RANGE CORRELATION OF THE MODEL AND HANG SENG INDEX

In this section, detrended fluctuation analysis (DFA) method is applied on the lattice Sierpinski carpet percolation. The DFA is a technique used to estimate a scaling exponent from the behavior of the average fluctuation of a random variable around its local trend, for the details see [26]. The cumulative deviation of time series $\{x_t, = 1, \ldots, N\}$ is given by

$$Y(i) = \sum_{k=1}^{i} (x_k - \overline{x}), \quad \text{for } i = 1, \ldots, N.$$

$$(5.1)$$

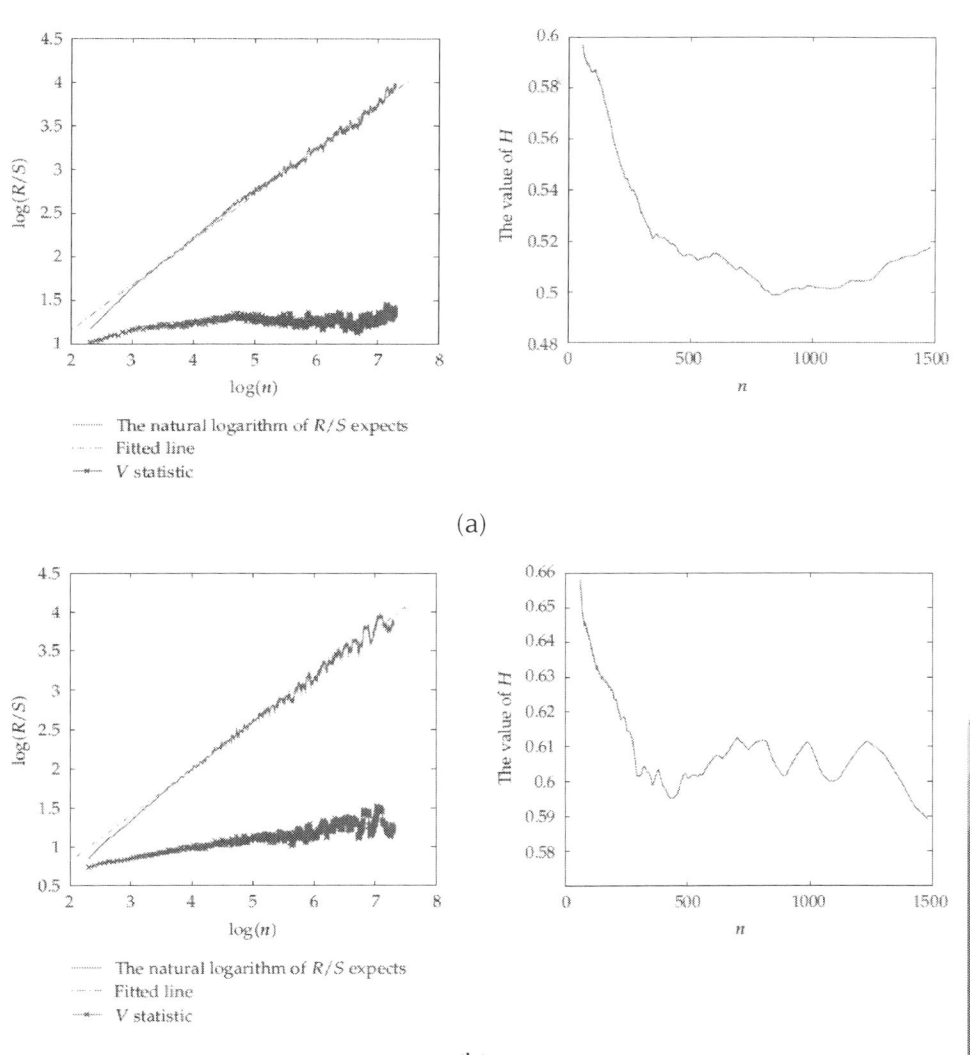

Figure 4: (a) The plots of modified R/S statistics and the fluctuation of exponent H for the price model. (b) The corresponding plots for the actual data from Hang Seng Index.

We divide (i) into intervals of nonoverlapping and equal length of time (n). Then the root mean square fluctuation for all such length interval is defined as

$$F(n) = \sqrt{\frac{1}{N} \sum_{i=1}^{N} [Y(i) - Y_n(i)]^2},$$

(5.2)

where $Y_n(i)$ is the fitting polynomial of the interval. The above definition is repeated for all the divided intervals. There is a power-law relation between (n) and n, namely,

$$F(n) \sim n^\alpha.$$

(5.3)

The parameter α is the scaling exponent or the correlation exponent, which exhibits the long-range correlation of the time series. For $\alpha=0.5$, it indicates that the time series is uncorrelated (white noise); for the value $0<\alpha<0.5$, it indicates the anticorrelations; for $0.5<\alpha<1$, the time series has the persistent long-range correlation. According to DFA method and computer simulation, the scaling exponents of the returns of the price model and Hang Seng Index are 0.51755 and 0.51531, respectively, in Figure 5. Although both the exponent values are larger than 0.5, they are very close to 0.5. This shows that there is some strong indication of long-range correlations for the returns.

CONCLUSIONS

A new random stock price model is developed by the lattice Sierpinski carpet percolation in the present paper, and a cluster of carpet percolation is applied to describe the cluster of traders sharing the same opinion about the market. The statistical properties of the returns are investigated and analyzed for different intensity values, and the behaviors of long memory and long-range correlation in volatility series are exhibited. Further, Hang Seng Index is also introduced and investigated by comparison; the empirical results show that the price model is accord with the real market to some degree.

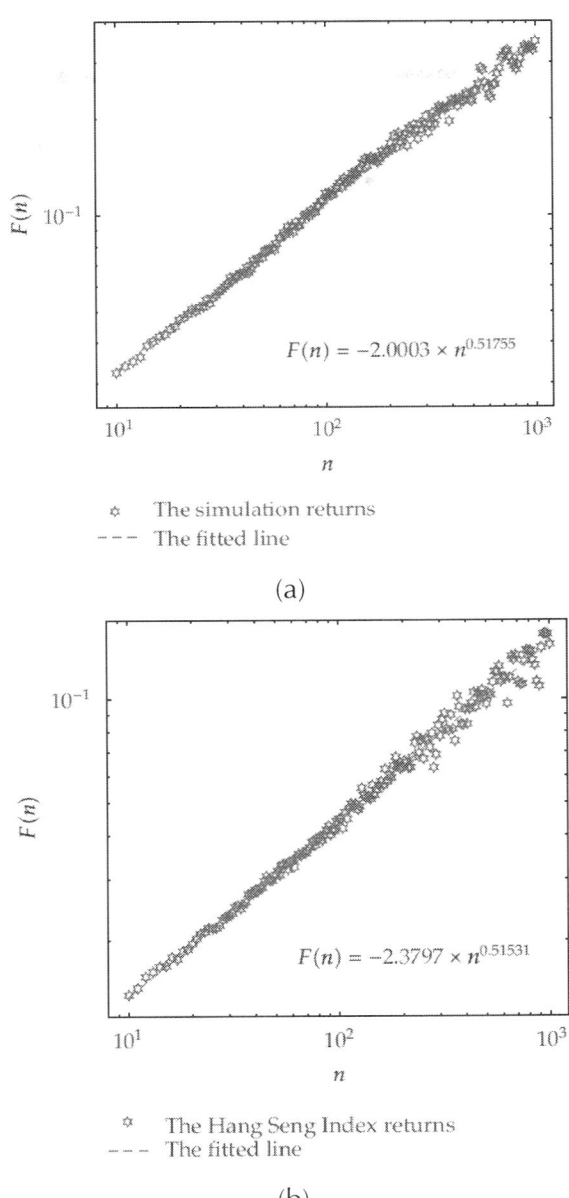

Figure 5: (a) DFA analysis of the returns for the simulation data with the intensity p=0.49. (b) DFA analysis of the returns for Hang Seng Index.

ACKNOWLEDGMENTS

The authors were supported in part by National Natural Science Foundation of China Grants no. 70771006 and no. 10971010, BJTU Foundation no. S11M00010.

REFERENCES

1. M.-F. Chen, From Markov Chains to Non-Equilibrium Particle Systems, World Scientific Publishing, River Edge, NJ, USA, 1992.

2. R. Durrett, Lecture Notes on Particle Systems and Percolation, Wadsworth & Brooks, 1998.

3. G. Grimmett, Percolation, vol. 321, Springer, Berlin, Germany, 2nd edition, 1999.

4. T. M. Liggett, Stochastic Interacting Systems: Contact, Voter and Exclusion Processes, vol. 324, Springer, New York, NY, USA, 1999.

5. T. M. Liggett, Interacting Particle Systems, vol. 276, Springer, New York, NY, USA, 1985.

6. D. Stauffer and A. Aharony, Introduction to Percolation Theory, Taylor & Francis, London, UK, 2001.

7. P. Bak, M. Paczuski, and M. Shubik, "Price variations in a stock market with many agents," Physica A, vol. 246, no. 3-4, pp. 430–453, 1997. · ·

8. R. Cont and J. P. Bouchaud, "Herd behavior and aggregate fluctuations in financial markets,"Macroeconomic Dynamics, vol. 4, no. 2, pp. 170–196, 2000. · ·

9. G. Iori, "Avalanche dynamics and trading fRICTIon effects on stock market returns," International Journal of Modern Physics C, vol. 10, no. 6, pp. 1149–1162, 1999. · · ·

10. T. Kaizoji, S. Bornholdt, and Y. Fujiwara, "Dynamics of price and trading volume in a spin model of stock markets with heterogeneous agents," Physica A, vol. 316, no. 1–4, pp. 441–452, 2002. · · ·

11. T. Lux and M. Marchesi, "Scaling and criticality in a stochastic multi-agent model of a financial market," Nature, vol. 397, no. 6719, pp. 498–500, 1999. · ·

12. D. Stauffer and D. Sornette, "Self-organized percolation model for stock market fluctuations," Physica A, vol. 271, no. 3-4, pp. 496–506, 1999. · ·

13. H. Tanaka, "A percolation model of stock price fluctuations," Mathematical Economics, no. 1264, pp. 203–218, 2002.

14. J. Wang and S. Deng, "Fluctuations of interface statistical physics models applied to a stock market model," Nonlinear Analysis: Real World Applications, vol. 9, no. 2, pp. 718–723, 2008. · ·

15. J. Wang, Q. Wang, and J. Shao, "Fluctuations of stock price model by statistical physics systems,"Mathematical and Computer Modelling, vol. 51, no. 5-6, pp. 431–440, 2010. ·

16. B. B. Mandelbrot, The Fractal Geometry of Nature, W. H. Freeman, San Francisco, Calif, USA, 1982.

17. E. E. Peters, Fractal Market Analysis: Applying Chaos Theory to Investment and Economics, John Wiley & Sons, New York, NY, USA, 1994.

18. M. Shinoda, "Existence of phase transition of percolation on Sierpiński carpet lattices." Journal of Applied Probability, vol. 39, no. 1, pp. 1–10, 2002. ·

19. J. Wang, "Supercritical ising model on the lattice fractal — the Sierpinski carpet," Modern Physics Letters B, vol. 20, no. 8, pp. 409–414, 2006. · · ·

20. R. Gaylord and P. Wellin, Computer Simulations with Mathematica: Explorations in the Physical, Biological and Social Science, Springer, New York, NY, USA, 1995.

21. K. Ilinski, Physics of Finance: Gauge Modeling in Non-Equilibrium Pricing, John Wiley, New York, NY, USA, 2001.

22. D. Lamberton and B. Lapeyre, Introduction to Stochastic Calculus Applied to Finance, Chapman & Hall, London, UK, 2000.

23. S. M. Ross, An Introduction to Mathematical Finance, Cambridge University Press, Cambridge, UK, 1999.

24. J. Wang, Stochastic Process and Its Application in Finance, Tsinghua University Press and Beijing Jiaotong University Press, Beijing, China, 2007.

25. F. Black and M. Scholes, "The pricing of options and corporate liabilities," Journal of Political Economy, vol. 81, pp. 637–654, 1973. · ·

26. P. Grau-Carles, "Long-range power-law correlations in stock returns," Physica A, vol. 299, no. 3-4, pp. 521–527, 2001. · · ·

27. Y. Guo and J. Wang, "Simulation and statistical analysis of market return fluctuation by Zipf method,"Mathematical Problems in Engineering, vol. 2011, Article ID 253523, 13 pages, 2011.

28. M. L›evy and S. Solomon, Microscopic Simulation of Financial Markets, Academic Press, New York, NY, USA, 2000.

29. D. Pirino, "Jump detection and long range dependence," Physica A, vol. 388, no. 7, pp. 1150–1156, 2009. · ·

30. Z. Zheng, Matlab Programming and the Applications, China Railway Publishing House, Beijing, China, 2003.

31. E. Zivot and J. H. Wang, Modeling Financial Time Series with S-PLUS, Springer, New York, NY, USA, 2006.

CITATION

Xu Wang and Jun Wang, "Statistical Behavior of a Financial Model by Lattice Fractal Sierpinski Carpet Percolation," Journal of Applied Mathematics, vol. 2012, Article ID 735068, 12 pages, 2012. doi:10.1155/2012/735068.

and the Support Vector Regression for Financial Time Series Forecasting

Jheng-Long Wu and Pei-Chann Chang
Department of Information Management,
Yuan Ze University, Taoyuan 32026, Taiwan

10

ABSTRACT

This paper presents a novel trend-based segmentation method (TBSM) and the support vector regression (SVR) for financial time series forecasting. The model is named as TBSM-SVR. Over the last decade, SVR has been a popular forecasting model for nonlinear time series problem. The general segmentation method, that is, the piecewise linear representation (PLR), has been applied to locate a set of trading points within a financial time series data However, owing to the dynamics in stock trading, PLR cannot reflect the trend changes within a specific time period. Therefore, a trend based segmentation method is developed in this research to overcome this issue. The model is tested using various stocks from America stock market with different trend tendencies. The experimental results show that the proposed model can generate more profits than other models. The model is very practical for real-world application, and it can be implemented in a real-time environment.

INTRODUCTION

Support vector machines (SVMs) have outperformed other forecasting models of machine learning or soft computing (SC) tools such as decision tree, neural network (NN), bayes classifier, fuzzy systems (FSs), evolutionary computation (EC), and chaos theory by many researchers from historical nonlinear time series data applications in the last decade [1–5]. In these techniques, many researchers presented different forecasting models in dealing with characteristics such as imprecision, uncertainty, partial truth, and approximation to achieve practicability, robustness, and low solution cost in real applications [6–8]. However, the most important issue in resolving the nonlinear time series problem is error revision. ANNs use the empirical risk minimization principle to minimize the generalization errors but SVRs use the structural risk minimization principle because SVR is able to analyze with small samples and to overcome the local optimal solution problem, which surpasses to ANNs [9–11]. Therefore, the SVRs forecasting model is applied to accomplish the forecasting task in this research. Presently, support vector regression (SVR), which was evolved from support vector machine (SVM) based on the statistical learning theory, is a powerful forecasting and machine learning approach for numerical prediction [12–15]. Also, SVR has high toleration error rate and high accuracy for learning solution knowledge in complex problems [16]. Although SVR can be applied well in time series data, the input vector is a key successful factor. Despite the volatile nature of the stock markets, researchers still can find certain correlations between these factors and stock prices. An investor's primary goal is to make profits. In order to help investors achieve their financial objectives, researchers have studied the relationship between financial markets and price variations over time from [17–20].

In the last few years, several representations of time series data have been proposed; the most often used representation is piecewise linear representation (PLR) [21–23]. It can decompose a time series data into a series of bottom and peak points [24, 25] in financial market. But the traditional PLR does not consider the multiple trending characteristics in time series. Moreover, the price movements of stocks are affected by many factors such as government policies, economic environments, interest rates, and inflation rates. The share prices of most listed companies also move up and down with other changing factors like market capitalization, earnings per share (EPS), price- to -earnings ratio, demand and supply, and market news. Moreover, there are more fractal properties of financial data, such as self-similarity, heavy-tailed distributions, long memory, as well as power laws [26–29]. One of fractal properties is long memory which is a common characteristic in financial data or other fields [30–32]. The daily stock trading is a short-term return so in this paper these fractal properties were not considered in our framework, just focusing on the real stock price's trends.

Therefore, there is a need to develop a new segmentation method which takes the price moving trends into consideration. As a result, this research will consider the multiple trends of stock price's movements in TBSM segmentation approach to capture the embedded knowledge of nonlinear time series. This research intends to improve the SVR forecasting performance using a trend based decomposition method. The TBSM approach has captured the tendency of stock price's movement which can be inputted into SVR in learning the historical knowledge of the time series data. Moreover, a more accurate forecasting result can be achieved when applied in real-time stock trading decision.

The rest of this paper is organized as follows. In Section 2, we describe TBSM segmentation principle. Forecasting model is discussed in Section 3. Section 4 explains modeling for trading decisions including using historical data to make trading decisions by the TBSM approach, selecting highly correlated technical indices by stepwise regression analysis (SRA), forecasting trading signals by SVR, and evaluating trading strategies. Section 5 explains how the TBSM with SVR for stock trading decisions and compares the profits obtained from various forecasting approaches. Finally, conclusions and directions for further research are discussed in Section 6.

A TREND BASED SEGMENTATION METHOD (TBSM)

In the time series database there are many approaches such as Fourier transform, wavelets, and piecewise linear representation which can be applied to find the turning point on time series data. According to the characteristics of sequential data, a piecewise linear representation of the data is more appropriate. A variety of algorithms to obtain a proper linear representation of segment data have been presented. As reported in [33–36], PLR is used to support more tasks and provides an efficient and effective solution. In this paper we intend to enhance the segmentation accuracy based on different trends in stock price's movements. The basic idea of TBSM is to modify the PLR segmentation using the trend tendency in a specific time period. Three different trends such as uptrend, downtrend, and hold trend will be considered when making the segmentation. Detailed procedures of TBSM include the following. (1) PLR is applied to locate the turning points from the time series including up or downtrends. (2) The points around each turning point will be double-checked if the variations of the points are within the threshold. If yes, these points will have the same buy/sell trading in this period. (3) These points are set to be in the same trend. The pseudocode of the TBSM is shown in Algorithm 1.

Define: Threshold // cutting threshold
X_Thld // horizontal area
Y_Thld // vertical area
X // a time series
Y // stock price
1: Procedure TBSM(T)
2: Let T be represented as X[1, 2,..., n], Y[1, 2,..., n]
3: n = 0
4: Draw a line between (X_1, Y_1) and (X_n, Y_n)
5: Max d = maximum distance of (X_i, Y_i) to the line
6: If (Max d > Threshold)
7: Let (X_i, Y_i) be the point with maximum distance
8: For j = X_1:X_n
9: If ($\left\|X_j - X_i\right\|$ < X_Thld) and ($\left\|Y_j - Y_i\right\|$ < Y_Thld)

10:	Then Point[n]=[X_i,X_i], n = n + 1
11:	End If
12:	End For
13:	Select from Point[n]:X_{t1}=Min(X_0),X_{t2}=Max(X_n)
14:	Return: S1 = T [X_1,X_{t1}]
15:	S2 = T [X_{t2},X_n]
16:	End If

Algorithm 1: A pseudocode for TBSM in time series data.

For example, a time series $T=\{t_1, t_2,...,t_{191}\}$ with 191 data is given to explain the basic idea of the TBSM procedure. As shown in Figure 1(a), several trading points are represented as buy (four red points) or sell (six green points) in this case. According to the TBSM procedure, we can draw a line S_1 form the first point to the last point as shown in Figure 1(b) and find the max distance to line S_1 which is point t_{26}. Then

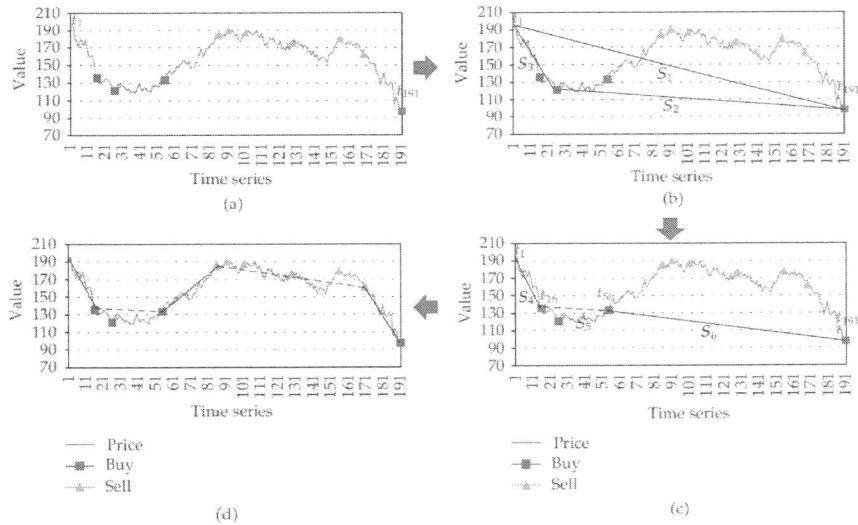

Figure 1: An example for TBSM in time series data.

line S_1 is decomposed into two segments including line S_2 from t_1 to t_{26} and line S_3 from t_{26} to $t_{19}1$. Based on point t_{26}, we can locate point t_{16} to t_{56} which are varied within the threshold. These points are set as hold trend and with the same state of point t_{26}. Therefore line S_2 and line S_3 will be changed to three different lines including line S_4 from point t_1 to point t_{16}, line S_5 from point t_{16} to point t_{56}, and line S_6 is from point t_{56} to point t_{191} as shown in Figure 1(c). Next step is repeating the same process for the rest of segments as t_{56} to t_{191}. The final results are shown in Figure 1(d) including two hold trend segments (dotted line), one uptrend segment, and two downtrend segments (solid line) in this time series.

SUPPORT VECTOR REGRESSIONS (SVRS)

Support vector regression is a modification of machine-learning-theory-based classification called support vector machine. Machine learning techniques have been applied for assigning trading signal. Many studies used support vector machine for determining whether a case contains particular class [37, 38]. But the shortcoming only deal with discrete class labels, whereas trading signal continuum data type because a weight of signal can take a buy or sell power. Grounded in statistical learning theory [1, 2], support vector regression is capable to predict the continuous trading signal while still benefiting from the robustness of SVM. SVM has been successfully employed to solve forecasting problems in many fields, such as financial time series forecasting [39] and emotion computation [40]. For explaining the concept of SVR, we have considered a standard regression problem. Let $S=\{X_i, Y_i\}$ $_{i=1...n}$ be the set of data where Xi is input vector (selected technical index in this research), Y_i (trading signal t_s) is an output vector, and n is the number of data points. In regression analysis, we find a function $f(X_i)$ such that $Y_i=f(X_i)$. This function can be used to find the output value Y of any X. The standard regression function is as follows:

$$q_i = f(x_i) + \delta,$$

(3.1)

where δ denotes the random error and q_i denotes the estimated output. There are two types of regression problems, namely, linear and nonlinear. SVR is developed to tackle the nonlinear regression problems because the nonlinear regression problems have high complexity as well as stock market trade. In SVR, at first the input vectors are non-linearly mapped into a high-dimensional feature space (F), where they are linearly correlated with the respective output values.

SVR uses the following linear estimation function:

$$f(x) = (\omega \cdot \phi(x)) + b,$$

(3.2)

where ω denotes the weight vector, b denotes a constant, $\phi(x)$ denotes the mapping function in the feature space, $(\omega \cdot \phi(x))$ and denotes the dot product in the feature space F SVR transfers the nonlinear regression problem of the lower dimension input space (x) into a linear regression problem of a high-dimension feature space. In other words, the optimization problem involving a nonlinear regression is converted into finding the flattest function in the feature space instead of input space.

Various cost functions like Laplacian, Huber's Gaussian, and ε insensitive can be used in the formulation of SVR. The cost function should be suitable for the problem and should not be very complicated because a complicated cost function could lead to difficult optimization problems. Thus, we have used robust ε-sensitive cost function which is shown below:

$$L_\varepsilon(f(x), q) = \begin{cases} |f(x) - q| - \varepsilon, & \text{if } |f(x) - q| \geq 0 \\ 0, & \text{otherwise,} \end{cases}$$

(3.3)

where ε denotes a precision parameter which represents the radius of the tube located around the regression function.f(x).

The {+ε,-ε}region is called ε insensitive zone. ε is determined by the user. If the actual output value lies in this region, the forecasting error is considered to be zero.

The weight vector, ω, and constant, b, in (3.2) are calculated by minimizing regularized risk function which is shown in (3.4):

$$R(C) = \frac{C}{n} \sum_{i=1}^{n} L_\varepsilon(f(x_i), q_i) + \frac{1}{2}|\omega|^2,$$

(3.4)

where $L_\varepsilon(f(x_i), q_i)$ denotes the -insensitive loss function, $|\omega^2|/2$ denotes the regularization term, and denotes the regularization constant. ω decides the complexity and approximate accuracy of the regression model. Value of C is selected by the user to ensure appropriate value of w and low empirical risk.

The two positive slack variables ξ_i and ξ_i^* are used to replace the -insensitive loss function of (3.3). ξ_i is defined as the distance between the qi and higher boundary of the ε insensitive zone, and ξ_i^* is defined as the distance between the qi and lower boundary of the -insensitive zone. Equation (3.4) is transformed into (3.5) by using the slack variables:

$$\text{Minimize}: R_{\text{reg}}(f) = \frac{1}{2}|\omega|^2 + C\sum_{i=1}^{n}(\xi_i + \xi_i^*)$$

(3.5)

$$\text{Subject to} \begin{cases} q_i - (\omega \cdot \phi(x_i)) - b \leq \varepsilon + \xi_i \\ (\omega \cdot \phi(x_i)) + b - q_i \leq \varepsilon + \xi_i^* \\ \xi_i, \xi_i^* \geq 0, \quad \text{for } i = 1, \ldots, n. \end{cases}$$

(3.6)

Lagrange function method is used to find the solution which minimizes the regression risk of (3.4) with the cost function in (3.3) which results in the following quadratic programming problem (QP):

$$\text{Minimize}: \frac{1}{2}\sum_{i=1}^{N}\sum_{j=1}^{N}(\alpha_i - \alpha_i^*)(\alpha_j - \alpha_j^*)(\phi(x_i) \cdot \phi(x_j))$$

$$+ \sum_{i=1}^{N}(\varepsilon_i^{\text{up}} - y_i)\alpha_i + \sum_{i=1}^{N}(\varepsilon_i^{\text{down}} - y_i)\alpha_i^*,$$

(3.7)

$$\text{Subject to}: \sum_{i=1}^{N}(\alpha_i - \alpha_i^*) = 0, \quad \text{where } \alpha_i, \alpha_i^* \in [0, C],$$

(3.8)

where α_i and α_i^* denote Lagrange multipliers. $\varepsilon_i^{\text{up}}$ and $\varepsilon_i^{\text{down}}$ represent the i th up- and downmargin, respectively. The value of $\varepsilon_i^{\text{up}}$ and $\varepsilon_i^{\text{down}}$ is equal to ε. The QP problem of (3.7) is solved under the constraints of (3.8). After solving the QP problem, we obtained Lagrange multiplier from (3.9), and (3.2) is transformed into the following equation (3.10):

$$w = \sum_{i=1}^{N} (\alpha_i - \alpha_i^*) \cdot \phi(x_i),$$

$$(3.9)$$

$$f(x) = (\alpha_i - \alpha_i^*)(\phi(x_i) \cdot \phi(x)) + b.$$

$$(3.10)$$

The Karush-Kuhn-Tucker (KKT) conditions are used to find the value of b KKT conditions state that at the optimal solution, the product between the Lagrange multipliers and the constraints is equal to zero. The value of b can be calculated as follows:

$$b = \begin{cases} y_i - (w \cdot \phi(x_i)) - \varepsilon_i^{\text{up}}, & \text{for } \alpha_i \in (0, C), \\ y_i - (w \cdot \phi(x_i)) + \varepsilon_i^{\text{down}}, & \text{for } \alpha_i^* \in (0, C). \end{cases}$$

$$(3.11)$$

Using the trick of the kernel function, (3.10) can be written as (3.12):

$$f(x) = \sum_{i=1}^{n} (\alpha_i - \alpha_i^*) K(x, x_i) + b,$$

$$(3.12)$$

where $K(x, x_i) = (\phi(x) \cdot \phi(x_i))$ denotes the kernel function which is symmetric and satisfies the Mercer's condition. SVR was able to predict the nonlinear relationship between technical indices and trading signal ts better than other soft computing (SC) techniques.

APPLICATION IN FINANCIAL TIME SERIES DATA

This paper proposes a forecasting framework using a TBSM combined with SVR model which is called TBSM-SVR trading model for stock trading. The framework of TBSM-SVR trading model has five stages: the first is generating nonlinear trading segments by TBSM approach from historical stock price; the second is trading signal transformation from trading segments; the third is feature selection from technical indices by SRA approach; the fourth is learning the trading forecasting model by SVRs approach. The framework of TBSM-SVR model is shown in Figure 2. The five stages of TBSM-SVR model are explained as follows.

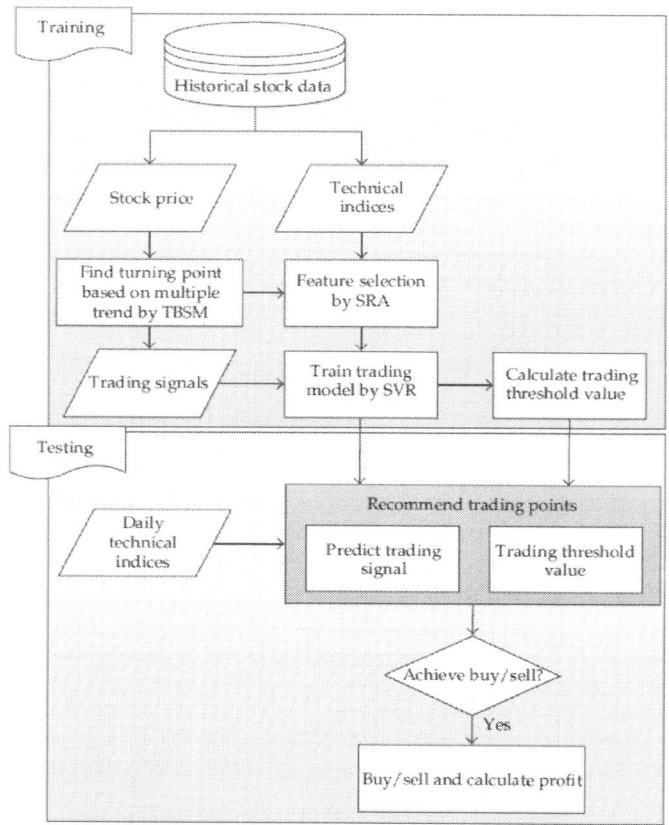

Figure 2: The framework of TBSM-SVR model for stock trading.

Find Turning Points Based on Multiple Trend by TBSM

According to TBSM procedure to find turning point based on trend of stock price, we selected a time series of historical stock price in a period to segment into several segments based on three trends including uptrend, downtrend, and hold trend. For example, a time series is given to segment trend segments from the date 2008/1/2 to 2008/12/30. Figure 3 shows the segmentation result by our proposed TBSM approach. The blue line is original historical stock price. The dashed lines are up/down a trend which if the segment trend goes up is belonging to uptrend and if the segment trend goes down is belonging to downtrend. The dot line is belonging to hold trend. In our experiment, each stock price can split to multiple trend segments for trading signal transformation.

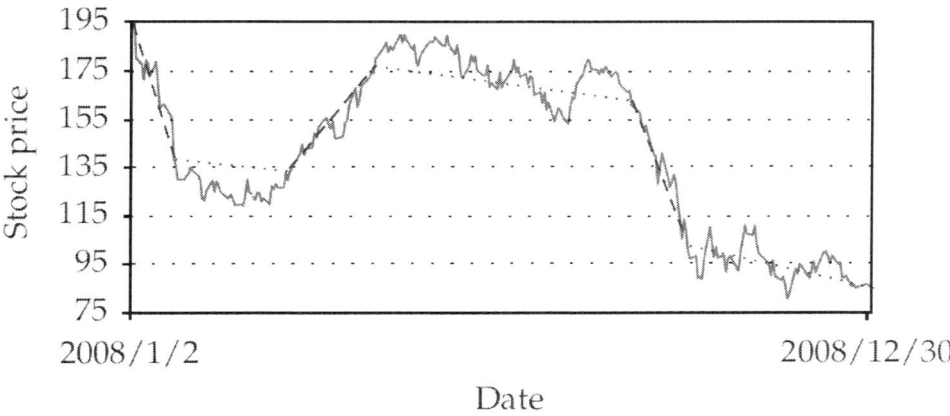

Figure 3: An example of segmentation result by TBSM.

Trading Signal Transformation

In this stage, the aim is calculating the trading signal for a nonlinear time series of segmentation result which are a lot of segments based on trends. We suppose a segment S_k is uptrend; then we assume the real value into the vector S'_k like to $S_k = [0, 0.1,..., 1]$; if S_k is hold trend but locates in buy point, then the vector like to $S'_k = [0.5, 0, 0.5]$; if S_k is hold trend but locates in sell point; then the vector like to $S'_k = [0.5, 1, 0.5]$; if S_k is downtrend, then the vector S'_k like to $[1, 0.9,..., 0]$. Finally we combine these S'_k to a full time series of trading signal *ts*. If the segment belongs to uptrend or downtrend, then the formula equation (4.1) is used to calculate trading signal value:

$$S'_{k,i} = \begin{cases} \dfrac{i}{L} & \text{if } S_k \text{ is uptrend segment,} \\ \dfrac{(L-i)}{L} & \text{if } S_k \text{ is downtrend segment,} \end{cases} \qquad (4.1)$$

where L denotes the length of segment S_k, whereas segment belonging to hold trend is using (4.2) to calculation:

$$S'_{k,i} = \begin{cases} 1 & \text{if } i\text{th is higherpoint in time series,} \\ 0 & \text{if } i\text{th is lower point in time series,} \\ 0.5 & \text{otherwise.} \end{cases} \quad (4.2)$$

For example, the S1, and S3 are hold trend; the S1 is down-trend; the S4 is up-trend. The result of trading signal ts is shown in Figure 4. The red dotted line is the hold trend which is a special signal for increasing reflects on the original turning points, so the hold trend is not a horizontal line. The purple dotted line is downtrend signal, and the orange dotted line is uptrend signal. For example, in the time series T the T1 to T5 and T10 to T14 are hold trend signal representation, T6 to T9 is downtrend signal representation, and finally T15 to T18 is uptrend signal representation. Finally the trading signal ts which is like to ts = {S1, S2, S3, S4} = { $\langle 0.5,0.5,1,0.5,0.5 \rangle$, $\langle 1,0.66,0.333,0 \rangle$, $\langle 0.5,0.5,0,0.5,0.5 \rangle$, $\langle 0,0.33,0.66,1 \rangle$ }. For the detail process see the pseudocode in Algorithm 2.

Input: length, oldTs // input data length and vector.
Output: newTs // a new time series vector of trading signal.
Method:
1:   Start = oldTs 1
2:   End = oldTs[length]
3:   If Start = = −1 and End = = 1
4:      newTs 1 = 0
5:      For i = 1: length−1
6:         newTs[i+1] = 1/(lenghth−1)*i
7:      End For
8:   Else If Start = = 1 and End = = −1
9:      newTs[length] = 0
10:      For i = 1 : length−1
11:         newTs[i+1] = 1/(lenghth−1)*(length−i)
12:      End For
13: Else
14:      For i = 2 : length−1
15:         newTs[i] = 0.5
16:      End For
17: End If

Algorithm 2: A pseudocode for trend segments by TBSM in time series.

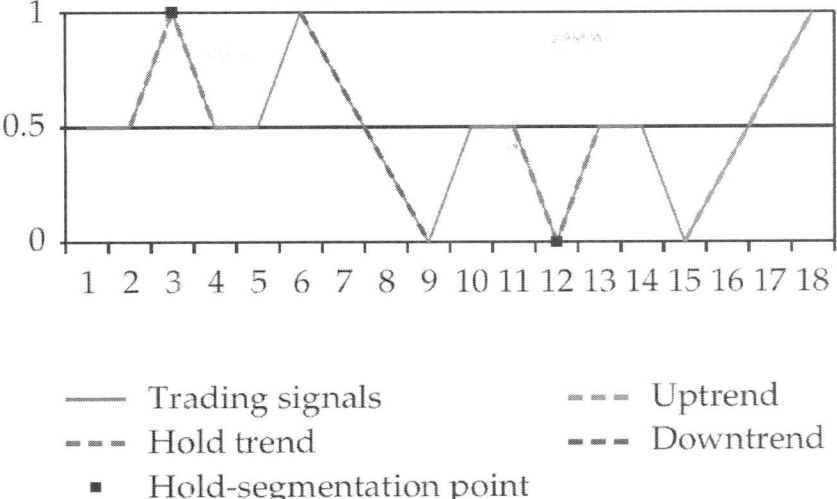

Figure 4: A sample of trading signal.

Feature Selection for Technical Indices by SRA

In this paper, we have considered 28 variables (technical indices) as listed in Table 1. These variables are correlated with variations in stock prices to some degree. The quantity of correlation varies for different variables. Rather than using all the 28 variables, we select the variables with a greater correlation than a user-defined threshold. The variable selection is done by stepwise regression analysis. We apply the SRA approach to determine which technical indices affecting the stock price. This is accomplished by selecting the variables repeatedly.

Table 1: Technical indices used as input variables

Technical	Technical index	Explanation
Moving average (Ma)	5 MA, 6 MA, 10 MA, 20 MA	Moving averages are used to emphasize the direction of a trend and smooth out price and volume fluctuations that can confuse interpretation.
Bias (BIAS)	5 BIAS, 10 BIAS	The difference between the closing value and moving average line, which uses the stock price nature of returning back to average price to analyze the stock market.
Relative strength index (RSI)	6 RSI, 12 RSI	RSI compares the magnitude of recent gains to recent losses in an attempt to determine overbought and oversold conditions of an asset.
Nine days stochastic line (K, D)	9 K, 9 D	The stochastic line K and line D are used to determine the signals of overpurchasing, overselling, or deviation.
Moving average convergence and divergence (MACD)	9 MACD	MACD shows the difference between a fast and slow exponential moving average (EMA) of closing prices. Fast means a short-period average, and slow means a long period one.
Williams %R (pronounced "percent R")	12 W%R	Williams %R is usually plotted using negative values. For the purpose of analysis and discussion, simply ignore the negative symbols. It is best to wait for the security's price to change direction before placing your trades.
Moving average convergence and divergence (MACD)	9 MACD	MACD shows the difference between a fast and slow exponential moving average (EMA) of closing prices. Fast means a short-period average, and slow means a long period one.
Williams %R (pronounced "percent R")	12 W%R	Williams %R is usually plotted using negative values. For the purpose of analysis and discussion, simply ignore the negative symbols. It is best to wait for the security's price to change direction before placing your trades.
Transaction volume (TV)	5 TV, 10 TV, 15 TV	Transaction volume is a basic yet very important element of market timing strategy. Volume provides clues as to the intensity of a given price move.
Differences of technical index (Δ)	Δ 5 MA, Δ 6 MA, Δ 10 MA, Δ 5 BIAS, Δ 10 BIAS, Δ 6 RSI, Δ 12 RSI, Δ 12 W%R, Δ 9 K, Δ 9 D, Δ 9 MACD	Differences of technical index between the day and next day.

In the feature selection part input factors will be further selected using stepwise regression analysis (SRA). The SRA has been applied to determine the set of independent variables which is most closely affecting the dependent variable. The SRA is step by step to select factor into regression model which if factor has the significance level, then it is selected. We can follow (4.4) to calculate the F value of SRA:

$$SSR = \sum \left(\hat{Y} - \overline{Y}\right)^2,$$

$$SSE = \sum \left(\hat{Y}_i - Y_i\right)^2, \tag{4.3}$$

$$F_j^* = \frac{MSR(x_j \mid x_i)}{MSE(x_j \mid x_i)} = \frac{SSR(x_j \mid x_i)}{SSE/(n-2)\ (x_j \mid x_i)} \quad i \in I, \tag{4.4}$$

where SSR denotes a regression sum of square. SSE denotes residual sum of squares. x is the value of technical index. y is the value of stock price. n is the total number of training data. $\hat{Y}$ is the forecasting value of regression. $\overline{Y}$ is the average stock price of training data. After the feature selection by SRA, we can provide a set of features to form an input vector for the next step to learning the forecasting model.

The steps of the SRA approach are described as follows.

Step 1. Find the correlation coefficient r for each technical index $\upsilon_1, \upsilon_2, ..., \upsilon_n$ with the stock price y in a stock. These correlation coefficients are stored in a matrix called correlation matrix.

Step 2. The technical index with largest R^2 value is selected from the correlation matrix. Let the technical index be υ_i. Derive a regression model between the stock price and technical index, that is $\hat{y} = f(\upsilon_i)$.

Step 3. Calculate the partial F value of other technical indices. Compare the R^2 value of the remaining technical indices and select the technical index with the highest correlation coefficient. Let the technical index be υ_j. Derive another regression model, that is $\hat{y} = f(\upsilon_i, \upsilon_j)$.

Step 4. Calculate the partial F value of the original data for the technical index υ_j. If the F value is smaller than the user-defined threshold, υ_j is removed from the regression model since it does not affect the stock price significantly.

Step 5. Repeat Step 3 to Step 4. If the F value of variable is more than the user-defined threshold, the variable should be added to the model, otherwise it should be removed.

In addition, the range of the input variables of SVR model should be between 0 and 1. Hence, the selected technical indices are normalized as follows:

$$\text{Normal}(x_{ij}) = \frac{x_{ij} - \text{Min}(x_i)}{\text{Max}(x_i) - \text{Min}(x_i)} \quad i = 1,\ldots,n; \ j = 1,\ldots,m; \ n, \ m \in \Re, \tag{4.5}$$

where Normal (x_{ij}) denotes the normalized value of jth data point of ith technical index. Max (x_i) denotes the maximum value of ith technical index. Min (x_i) denotes the minimum value of ith technical index. x_{ij} denotes original value of jth data point of ith technical index. n and m denote the total number of technical indices and data points, respectively

Learning the Trading Forecasting Model by SVR

Support vector regression will be applied as a machine learning model to extract the hidden knowledge in the historic stock database. The single output is the trading signal ts from TSBM process, and the multiple input features are technical indices from SRA selection. SVR learning model transforms multiple features into high multidimensional feature space, and the transformed feature space can be mapped into a hyperplane space to determine correct signals based on those support vector points. On the kernel function selection, we try to use linear, RBF, polynomial, and sigmoid functions to generate better performance for the SVR model because the stock market is a very complicated nonlinear environment. Since the SVR approach possesses high learning capability and accuracy in predicting continuous signals for building hidden knowledge among trading signals and technical indices, it is a widely used tool for predicting the trading signals.

Trading Points Decision from Forecasted Trading Signal

In the daily forecasting, if the forecasted trading signals by SVR satisfied buy threshold, then this means it is needed to buy stock quickly because it is very close to turning point; otherwise if the state satisfied a sell threshold, then there is need to sell stock. These satisfied points are recommended to transaction in stock market. Before determining the trading point, we will calculate the buy/sell threshold values for two trading types. The trading thresholds of two types are as follows:

$$\text{Buy}_{\text{threshold}} = \mu + \sigma,$$

$$\text{Sell}_{\text{threshold}} = 1 - \mu + \sigma,$$

$$\mu = \frac{1}{N}\sum_{i=1}^{N}x_i',$$

$$\sigma = \sqrt{\frac{1}{N}\sum_{i=1}^{N}(x_i' - \mu)}, \tag{4.6}$$

where μ denotes the average of trading signal in training data. σ denotes the standard deviation of trading signal in training data. Buy_threshold denotes the buy trading threshold. Sell_threshold denotes the sell trading threshold. If forecasted trading signals form SVR model in testing data are more than buy_threshold, then this suggests trading point for buy stocks else if forecasting signal in testing data is smaller than sell_threshold, then this suggests trading for sell stock.

In the trading decision step, the TBSM-SVR model is employed to calculate daily trading signals. The detailed principles for making trading decisions include the following.

1. If the time series prediction of trading signals by TBSM-SVR model is going up and intersects with buy trading threshold Buy_threshold, then it is a "buy" trading decision.

2. If the time series prediction of trading signals by TBSM-SVR model is going down and intersects with sell trading threshold sell_threshold, then it is a "sell" trading decision.

3. A "hold" trading decision is made (or do not make any trading decision) when the forecasting trading signal does not intersect with buy and sell thresholds.

For example, Figure 5 shows trading points decision for Apple stock. How to suggest the buy/sell points for stock in a time series in which the red square points are buy points and green triangle points are the sell points? Both are satisfied two thresholds in which the orange dotted line is sell threshold and the purple dotted line is buy threshold, so we can forecast the trading points daily by an automatically trading system.

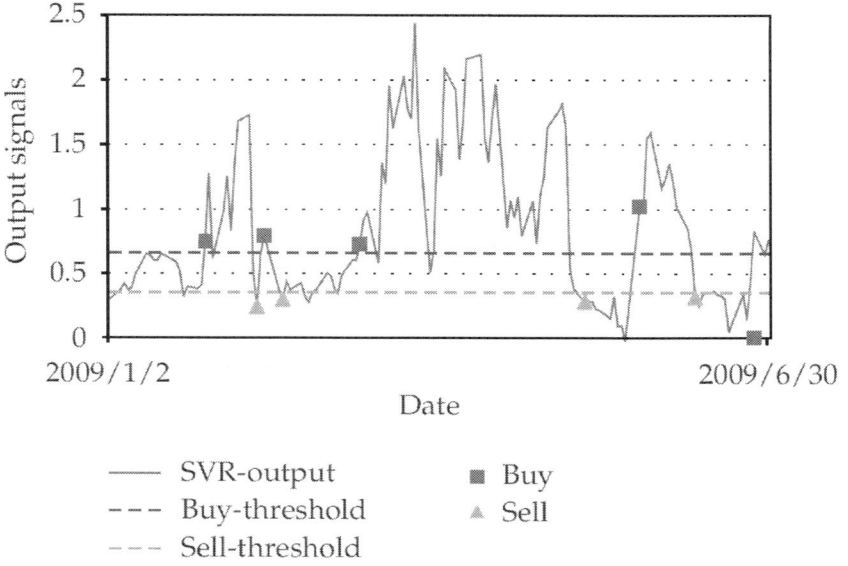

Figure 5: An example of result for detecting trading points of Apple.

EXPERIMENTAL RESULTS

Profit Evaluation and Parameters Setting

In this research, the trading point (buy and sell timing) is decided by the TBSM-SVR model based on the forecasting trading signal of SVR and TBSM segmentation. In the experimental section, we also use various forecasting models to the generated profiting trading points and compare their performances. The profits in each different forecasting model are calculated as follows:

$$\text{profits} = C \prod_{i=1}^{k} \left\{ \frac{(1 - a - b) \times p_{S_i} - (1 + a) \times p_{B_i}}{(1 + a) \times p_{B_i}} \right\}, \tag{5.1}$$

where C is the total amount of money to be invested at the beginning as well as the capital of money, a refers to the tax rate of ith transaction, b refers to the handling charge of ith transaction, k is the total number of transaction, p_{Si} is the selling price of the ith transaction and p_{Bi} is the buying price of ith transaction.

This study uses minimal root mean square error (RMSE) to measure the model performance in SVR train stage. In the model selection strategy that the dataset uses the last one trading period of training data contains (buy/sell and sell/buy states). The RMSE of an estimator $\hat{ts}$ with respect to the estimated parameter ts is defined as the square root of the mean square error:

$$\text{RMSE} = \sqrt{\frac{\sum_{i=1}^{n} ts_i - \widehat{ts_i}}{N}}.$$

(5.2)

ts denotes the trading signal by trading signal transformation from TBSM segmentation in Section 4.2. $\hat{ts}$ denotes the estimated trading signal by SVR forecasting model. N denotes total number in each training data (Table 2).

Table 2: The parameter setup for TBSM and SVR by DOEs (design of experiments)

Approach	Parameter	Value	Explanation
TBSM	Threshold	$0.1\ \sigma$ to $1\ \sigma$	The difference of price at uptrend or downtrend
TBSM	X_Thld	$0.1\ \sigma$ to $1\ \sigma$	The difference of days at hold trend
TBSM	Y_Thld	$0.1\ \sigma$ to $1\ \sigma$	The difference of price at hold trend
SVR	C	^{3}to^3	Cost
SVR	ε	^{4}to	Epsilon
SVR	d	^{9}to	Degree
SVR	g	2^1to2^4	Gamma

In parameter section we use design of experiments (DOEs) approach to set each parameter for capture optimal parameter combination for trading system in financial data. The parameters of the TBSM are based on standard deviation σ from stock price in each stock which is the range from $0.1\ \sigma$ to $1\ \sigma$ for testing in each parameters. In SVR model, the kernels chosen for testing are "radial basis function (RBF)" and "polynomial" function. The common combination includes cost

C;epsilon and are selected by the grid search with exponentially growing sequences. C ranges from 10^{-3} to 10^3. ε from 10^{-4} to 10^{-1} and γ is fixed as 0. In "polynomial" function, the degree d ranges from 2^{-9} to.2^{-1} The gamma g ranges from 2^1 to 2^4 in RBF kernel.

Profit Comparison in the US Stock Market

In this research, we have selected 7 stocks from the US stock market to compare the profit achieved by various trading models, including Apple, BOENING CO. (BA), Caterpillar Inc. (CAT), Johnson and Johnson (JNJ), Exxon Mobil Corp. (XOM), Verizon Communication Inc. (VZ), and S&P 500. Among all the stocks, 253 data points were collected for the training period from 1/2/2008 (mm/dd/yy) to 12/31/2008 while 124 data points were used for the testing period from 1/2/2009 to 6/30/2009. In this research, we have compared our forecasting model of TBMS-SVR approach with two other identification models developed in the past. The PLR-BPN model proposed by Chang et al. [26] used neural networks in combination with PLR and exponential smoothing to determine the trading points. Kwon and Kish [41] used statistical model such as moving average, rate of change and trading volumes to determine the buy-sell points and generated profit.

The technical indices selected result by SRA as shown in Table 3. Apple, Ba, CAT, JNJ, S&P 500, and VZ used 5 features (technical indices) for training forecasting model; XOM used 3 features for training forecasting model. From this result we can know that a few features can capture more trading knowledge.

Table 3: Feature selection result in each stock for technical indices by SRA

Stock	Technical index
Apple	5 MA, 6 MA, 9 K, 9 MACD, 12 W%R
BA	5 MA, 6 MA, 9 K, 10 TV, 12 W%R
CAT	5 MA, 6 MA, 9 K, 10 TV, 5 MA
JNJ	5 MA, 6 MA, 6 RSI, 9 MACD, 5 MA
S&P 500	5 MA, 5 BIAS, 10 TV, 26 BR, TAPI
VZ	5 MA, 6 MA, 5 MA, 10 TV, 26 VR
XOM	5 MA, 6 MA, 5 MA

From model selection results the RBF kernel has better low error in each stock by RMSE. Moreover, the gamma, degree, cost, epsilon, support vectors, and RMSE as shown in Table 4 are necessary parameters and measures. The models of TBSM-SVR in each stock are selecting optimal parameter combination by RMSE consideration.

Table 4: Model selection results from TSBM-SVR model for each stock

Stock	Kernel									
	Radial basis function (RBF)					Polynomial				
	g	C		SVs	RMSE	d	C		SVs	RMSE
Apple	2−1	103	10−4	253	0.0819	2	[0.001 : 1000]	[0.0001 : 0.1]	71	0.266
BA	2−1	103	10−1	107	0.0955	2	[0.001 : 1000]	[0.0001 : 0.1]	76	0.269
CAT	2−1	103	10−3	254	0.0898	2	[0.001 : 1000]	[0.0001 : 0.1]	156	0.233
JNJ	2−1	102	10−1	137	0.2617	1	[0.001 : 1000]	[0.0001 : 0.1]	116	0.426
S&P 500	2−1	103	10−4	254	0.0004	1	[0.001 : 1000]	[0.0001 : 0.1]	112	0.379
VZ	2−1	103	10−3	251	0.0031	1	[0.001 : 1000]	[0.0001 : 0.1]	125	0.269
XOM	2−1	103	10−4	253	0.0001	2	[0.001 : 1000]	[0.0001 : 0.1]	182	0.18

Each forecasting model provides trading points for each stock, so the best profits of the 3 forecasting models are shown in Table 5. The results turn out that our proposed TBSM with SVR model generates the greatest returns for the seven stocks, that is, number 1, 2, 3, 4, 5, 6, and 7 outperform other models. The average profit rate of these seven stocks is 40.42% using the TBSM-SVR model whereas the average profit rate generated by other models like PLR-SVR, PLR-BPN, and Statistical is 19.46%, 12.32%, and 9.65%, respectively. Therefore, our TBSM approach is better than PLR approach which is only considered linear representation.

Table 5: Comparison of profit obtained by various forecasting models

Stock no.	Stock name	TBSN-SVR model (RBF)	PLR-SVR model (RBF)	PLR-BPN model	Statistical model
1	Apple	92.35%	35.84%	12.97%	20.50%
2	BA	59.49%	35.69%	17.50%	20.03%
3	CAT	43.39%	36.09%	9.36%	24.83%
4	JNJ	13.95%	9.47%	16.88%	0%
5	S&P 500	22.78%	4.19%	3.77%	9.81%

6	VZ	28.60%	2.60%	27.72%	0%
7	XOM	22.40%	12.34%	−1.99%	−7.65%
Average		40.42%	19.46%	12.32%	9.65%

The buy and sell points obtained from the TBSM forecasting model in each stock are shown in Figures 6, 7, 8,9, 10, 11, and 12. The red square represents the buy point, and the black triangle represents the sell point using a trading strategy to determine turning points. Furthermore, our proposed approach TBSM is better than PLR segmentation which denotes that TBSM approach captures better trading knowledge for SVR forecasting model. Due to PLR only the linear representation is considering, so it loses important trend. Therefore, TBSM is an effective segmentation method for nonlinear time series data in stock market.

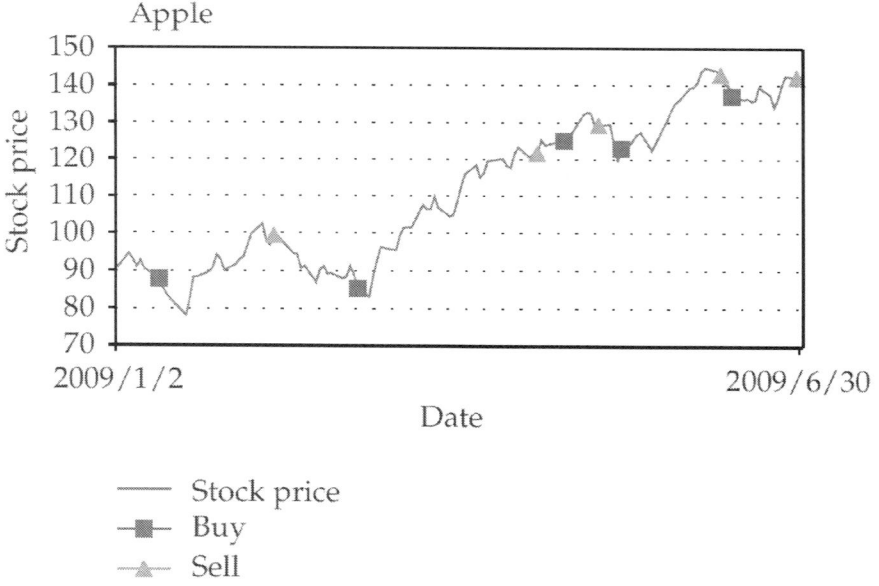

Figure 6: The forecasted trading points of Apple (an uptrend stock).

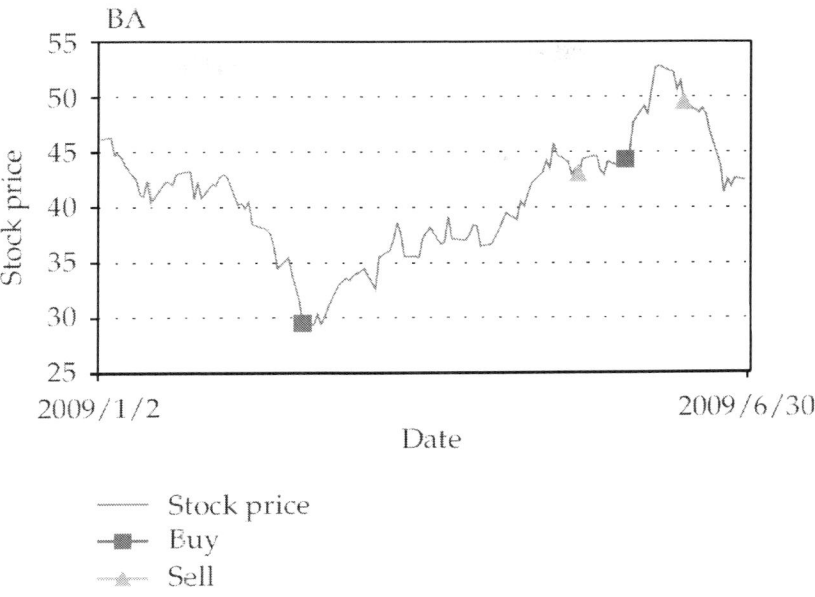

Figure 7: The forecasted trading points of BA (a steady-trend stock).

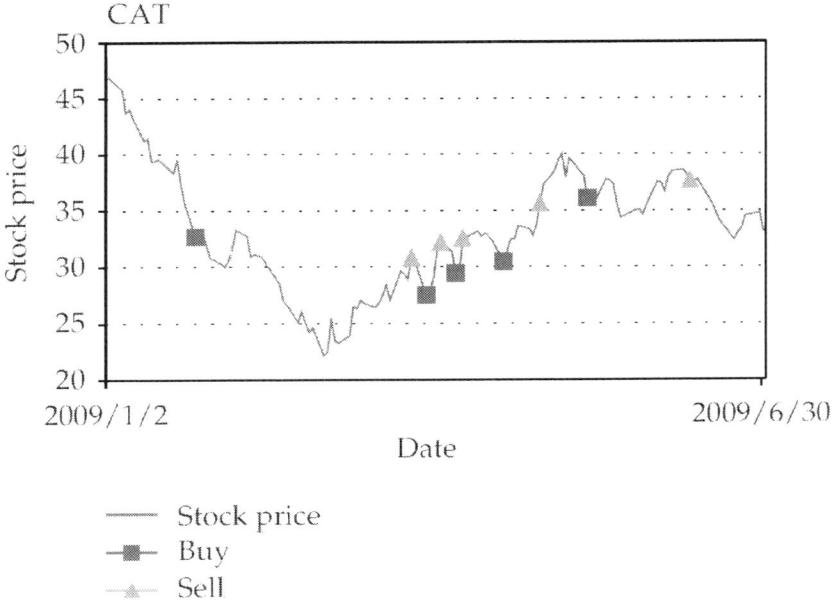

Figure 8: The forecasted trading points of CAT (a downtrend stock).

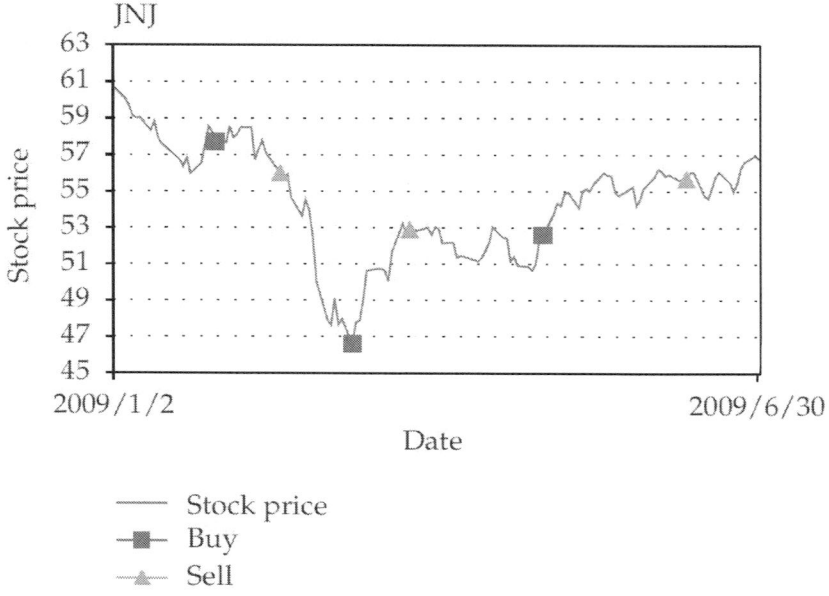

Figure 9: The forecasted trading points of JNJ (a steady-trend stock).

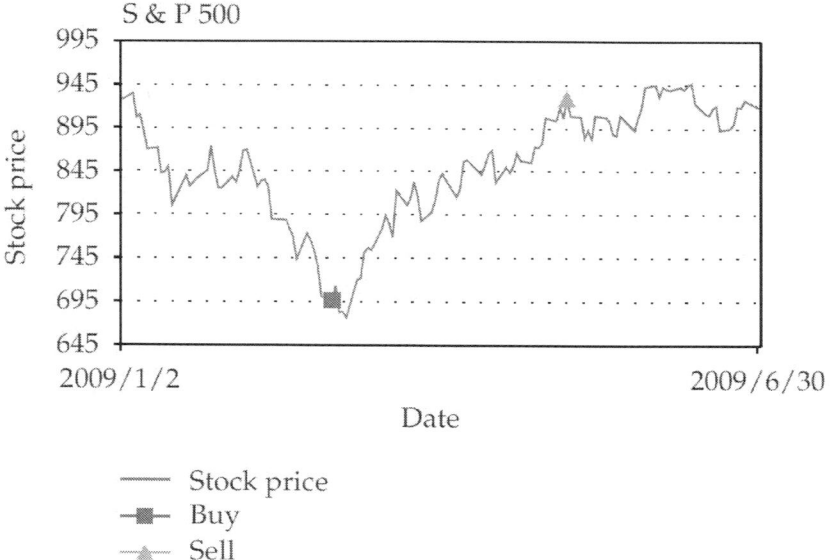

Figure 10: The forecasted trading points of S&P 500 (a steady-trend stock).

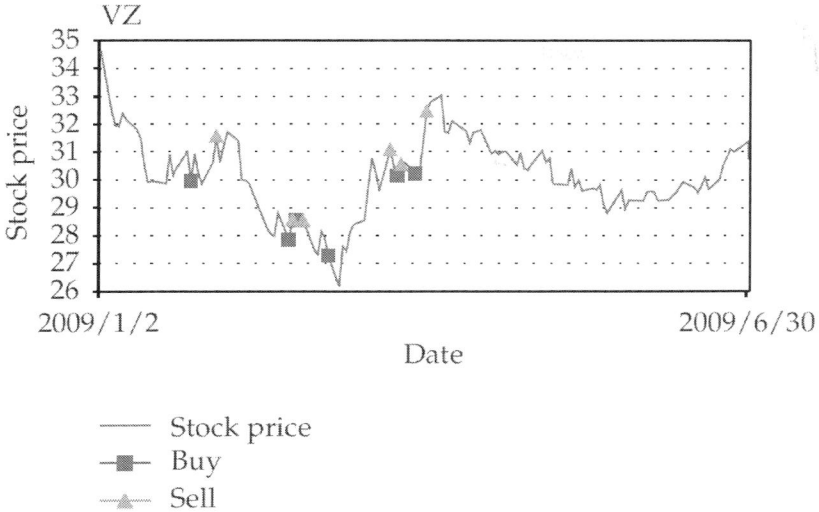

Figure 11: The forecasted trading points of VZ (a downtrend stock).

Figure 12: The forecasted trading points of XOM (a downtrend stock).

CONCLUSIONS

In this paper we proposed a trading system combining TBSM with SVR, and it is called TBSM-SVR-based stock trading system. This new trading system has been very effective in earning high profit while with the greatest ability. Experimental results showed that the TSBM can segment the stock price's variation into different trading trends. The trading signal in each trading trend will be assumed to be the same. The nonlinear time series can be better represented using these trading trends. Additionally, SVR is applied to capture the trading knowledge using the trading signals derived from these trading trends. The captured knowledge is more effective using TBSM-SVR when compared to PLR segmentation method. As a result, the primary goal of the investor could be easily achieved by providing him with simple trading decisions. However, the limitation of the TBSM-SVR trading system is the machine learning tool; that is, SVR is still not that mature yet. There are still rooms for the improvement of a better machine learning mechanism to be developed. Therefore, the trading system may make a wrong trading and lose money. In the future works, we can extend the segmentation method by considering a more detailed trend by investigating different buy-hold strategy or better trading strategy. In addition, the trend based segmentation method can further consider the fractal properties such as long memory, which can be accommodated to improve the segmentation performances.

REFERENCES

1. A. J. Smola and B. Schölkopf, "A tutorial on support vector regression," Statistics and Computing, vol. 14, no. 3, pp. 199–222, 2004. · ·

2. A. Swishchuk and R. Manca, "Modeling and pricing of variance and volatility swaps for local semi-markov volatilities in financial engineering," Mathematical Problems in Engineering, vol. 2010, Article ID 537571, 17 pages, 2010. · · ·

3. B. J. Chen, M. W. Chang, and C. J. Lin, "Load forecasting using support vector machines: a study on EUNITE Competition 2001," IEEE Transactions on Power Systems, vol. 19, no. 4, pp. 1821–1830, 2004. · ·

4. D. Niu, Y. Wang, and D. D. Wu, "Power load forecasting using support vector machine and ant colony optimization," Expert Systems with Applications, vol. 37, no. 3, pp. 2531–2539, 2010. ··

5. D. She and X. Yang, "A new adaptive local linear prediction method and its application in hydrological time series," Mathematical Problems in Engineering, vol. 2010, Article ID 205438, 15 pages, 2010. ··

6. E. Keogh and M. Pazzani, "An enhanced representation of time series which allows fast and accurate classification, clustering and relevance feedback," in Proceedings of the 4th International Conference on Knowledge Discovery and Data Mining (KDD '98), pp. 239–241, August 1998.

7. F. Girosi, M. Jones, and T. Poggio, "Regularization theory and neural networks architectures," Neural Computation, vol. 7, pp. 219–269, 1995.

8. F. X. Diebold and R. S. Mariano, "Comparing predictive accuracy," Journal of Business and Economic Statistics, vol. 20, no. 1, pp. 134–144, 2002. ··

9. H. Liu and J. Wang, "Integrating independent component analysis and principal component analysis with neural network to predict Chinese stock market," Mathematical Problems in Engineering, vol. 2011, Article ID 382659, 15 pages, 2011.

10. H. Wu, B. Salzberg, and D. Zhang, "Online event-driven subsequence matching over financial data streams," in Proceedings of the ACM SIGMOD International Conference on Management of Data (SIGMOD '04), pp. 23–34, June 2004.

11. J. L. Wu, L. C. Yu, and P. C. Chang, "Emotion classification by removal of the overlap from incremental association language features," Journal of the Chinese Institute of Engineers, vol. 34, no. 7, pp. 947–955, 2011.

12. J. O. Lachaud, A. Vialard, and F. De Vieilleville, "Analysis and comparative evaluation of discrete tangent estimators," in Proceedings of the 12th International Conference on Discrete Geometry for Computer Imagery (DGCI '05), E. Andres, G. Damiand, and P. Lienhardt, Eds., vol. 3429,, pp. 240–251, Springer, April 2005.

13. K. Y. Kwon and R. J. Kish, "Technical trading strategies and return predictability: NYSE," Applied Financial Economics, vol. 12, no. 9, pp. 639–653, 2002. ··

14. L. Muchnik, A. Bunde, and S. Havlin, "Long term memory in extreme returns of financial time series,"Physica A, vol. 388, no. 19, pp. 4145–4150, 2009. ··

15. L. Todorova and B. Vogt, "Power law distribution in high frequency financial data? An econometric analysis," Physica A, vol. 390, no. 23-24, pp. 4433–4444, 2011. ·

16. L. Zhang, W. D. Zhou, and P. C. Chang, "Generalized nonlinear discriminant analysis and its small sample size problems," Neurocomputing, vol. 74, no. 4, pp. 568–574, 2011. · ·

17. M. K. P. So, C. W. S. Chen, J. Y. Lee, and Y. P. Chang, "An empirical evaluation of fat-tailed distributions in modeling financial time series," Mathematics and Computers in Simulation, vol. 77, no. 1, pp. 96–108, 2008. · ·

18. M. Li and W. Zhao, "Visiting power laws in cyber-physical networking systems," Mathematical Problems in Engineering, vol. 2012, Article ID 302786, 13 pages, 2012.

19. M. Li, "Fractal time series—a tutorial review," Mathematical Problems in Engineering, vol. 2010, Article ID 157264, 26 pages, 2010. · ·

20. M. Li, C. Cattani, and S. Y. Chen, "Viewing sea level by a one-dimensional random function with long memory," Mathematical Problems in Engineering, vol. 2011, Article ID 654284, 13 pages, 2011. · ·

21. M. S. Abd-Elouahab, N. E. Hamri, and J. Wang, "Chaos control of a fractional-order financial system,"Mathematical Problems in Engineering, vol. 2010, Article ID 270646, 18 pages, 2010. · · ·

22. N. Muttil and K. W. Chau, "Neural network and genetic programming for modelling coastal algal blooms," International Journal of Environment and Pollution, vol. 28, no. 3-4, pp. 223–238, 2006. · ·

23. N. Sapankevych and R. Sankar, "Time series prediction using support vector machines: a survey," IEEE Computational Intelligence Magazine, vol. 4, no. 2, pp. 24–38, 2009. · ·

24. P. C. Chang and C. H. Liu, "A TSK type fuzzy rule based system for stock price prediction," Expert Systems with Applications, vol. 34, no. 1, pp. 135–144, 2008.

25. P. C. Chang, C. Y. Fan, and C. H. Liu, "Integrating a piecewise linear representation method and a neural network model for stock trading points prediction," IEEE Transactions on Systems, Man and Cybernetics Part C, vol. 39, no. 1, pp. 80–92, 2009. · ·

26. P. C. Chang, C. Y. Tsai, C. H. Huang, and C. Y. Fan, "Application of a case base reasoning based support vector machine for financial time series

data forecasting," in Proceedings of the International Conference on Intelligent Computing (ICIC '09), vol. 5755, pp. 294–304, September 2009.

27. P. F. Pai and C. S. Lin, "A hybrid ARIMA and support vector machines model in stock price forecasting," Omega, vol. 33, no. 6, pp. 497–505, 2005. · ·

28. P. F. Pai and W. C. Hong, "Forecasting regional electricity load based on recurrent support vector machines with genetic algorithms," Electric Power Systems Research, vol. 74, no. 3, pp. 417–425, 2005. · ·

29. S. Ghosh, P. Manimaran, and P. K. Panigrahi, "Characterizing multi-scale self-similar behavior and non-statistical properties of fluctuations in financial time series," Physica A, vol. 390, no. 23-24, pp. 4304–4316, 2011. · ·

30. S. O. Lozza, E. Angelelli, and A. Bianchi, "Financial applications of bi-variate Markov processes,"Mathematical Problems in Engineering, vol. 2011, Article ID 347604, 15 pages, 2011. ·

31. T. Farooq, A. Guergachi, and S. Krishnan, "Knowledge-based Green's Kernel for support vector regression," Mathematical Problems in Engineering, vol. 2010, Article ID 378652, 16 pages, 2010. · ·

32. V. Lavrenko, M. Schmill, D. Lawrie, P. Ogilvie, D. Jensen, and J. Allan, "Mining of concurrent text and time series," in Proceedings of the 6th International Conference on Knowledge Discovery and Data Mining (KDD '00), pp. 37–44, August 2000.

33. V. N. Vapnik, Statistical Learning Theory, Adaptive and Learning Systems for Signal Processing, Communications, and Control, John Wiley & Sons, New York, NY, USA, 1998.

34. V. N. Vapnik, The Nature of Statistical Learning Theory, Springer, New York, NY, USA, 1995.

35. W. C. Hong, "Chaotic particle swarm optimization algorithm in a support vector regression electric load forecasting model," Energy Conversion and Management, vol. 50, no. 1, pp. 105–117, 2009. · ·

36. X. H. Yang, D. X. She, Z. F. Yang, Q. H. Tang, and J. Q. Li, "Chaotic bayesian method based on multiple criteria decision making (MCDM) for forecasting nonlinear hydrological time series," International Journal of Nonlinear Sciences and Numerical Simulation, vol. 10, no. 11-12, pp. 1595–1610, 2009. ·

37. X. P. Ge, "Pattern matching in financial time series data," Computer Communications, vol. 27, pp. 935–945, 1998.

38. Y. W. Wang, P. C. Chang, C. Y. Fan, and C. H. Huang, "Database classification by integrating a case-based reasoning and support vector machine for induction," Journal of Circuits, Systems and Computers, vol. 19, no. 1, pp. 31–44, 2010. · ·

39. Y. Zhu, D. Wu, and S. Li, "A piecewise linear representation method of time series based on feature pints," in Proceedings of the11th International Conference on Knowledge-Based Intelligent Information and Engineering Systems (KES '07), 17th Italian Workshop on Neural Networks (WIRN '07), pp. 1066–1072, January 2007.

40. Z. Liu, "Chaotic time series analysis," Mathematical Problems in Engineering, vol. 2010, Article ID 720190, 31 pages, 2010. · ·

41. Z. Zhang, J. Jiang, X. Liu et al., "Pattern recognition in stock data based on a new segmentation algorithm," in Proceedings of the 2nd International Conference on Knowledge Science, Engineering and Management (KSEM '07), vol. 4798 of Lecture Notes in Computer Science, pp. 520–525, 2007.

CITATION

Jheng-Long Wu and Pei-Chann Chang, "A Trend-Based Segmentation Method and the Support Vector Regression for Financial Time Series Forecasting," Mathematical Problems in Engineering,. doi:10.1155/2012/615152.

A

B

D

G

K

M

N